OPTIMIZING THE OPERATION OF A MULTIPLE RESERVOIR SYSTEM IN THE EASTERN NILE BASIN CONSIDERING WATER AND SEDIMENT FLUXES

T0136225

Reem Fikri Mohamed Osman Digna

OPTIMIZING THE OPERATION OF A MULTIPLE RESERVOIR SYSTEM IN THE EASTERN NILE BASIN CONSIDERING WATER AND SEDIMENT FLUXES

DISSERTATION

Submitted in fulfillment of the requirements of
the Board for Doctorates of Delft University of Technology
and
of the Academic Board of the IHE Delft
Institute for Water Education
for
the Degree of DOCTOR
to be defended in public on
Tuesday 26 May 2020, at 15:00 hours
in Delft, the Netherlands

by

Reem Fikri Mohamed Osman DIGNA
Master of Science in Water Resources Engineering, University of Khartoum
born in Khartoum, Sudan

This dissertation has been approved by the
promotors: Prof.dr.ir. P. van der Zaag IHE Delft / TU Delft
            Prof.dr. S. Uhlenbrook IHE Delft / TU Delft
and
copromotor Dr. Y. Mohamed IHE Delft

Composition of the doctoral committee:

Rector Magnificus TU Delft   Chairman
Rector IHE Delft   Vice-Chairman

Prof. dr.ir. P. van der Zaag   IHE Delft / TU Delft, promotor
Prof. dr. S. Uhlenbrook   IHE Delft / TU Delft, promotor
Dr. Y. Mohamed   IHE Delft, copromotor

Independent members:
Prof.dr. D. P. Solomatine   IHE Delft / TU Delft
Prof.dr. G.P.W. Jewitt   IHE Delft / University of Kwazulu-Natal,
                                      South Africa
Prof.dr. E. Van Beek   University of Twente
Prof.dr. S. Hamad   NBI, Uganda
Prof.dr.ir. N.C. van de Giesen   TU Delft, reserve member

This research was conducted under the auspices of the SENSE Research School for
Socio-Economic and Natural Sciences of the Environment

CRC Press/Balkema is an imprint of the Taylor & Francis Group, an informa business

Published by:
CRC Press/Balkema
Schipholweg 107C, 2316 XC, Leiden, the Netherlands
Pub.NL@taylorandfrancis.com
www.crcpress.com – www.taylorandfrancis.com
ISBN 978-0-367-56441-4

ACKNOWLEDGEMENTS

First and foremost, I praise Allah for helping me to complete this study.

I am sincerely grateful to the high quality of supervision given by Prof. Stefan Uhlenbrook, Prof. Pieter van der Zaag and Prof. Yasir Mohamed. Prof. Stefan, I appreciate your continuous support you have given me during my PhD journey. Prof. Pieter, I am very grateful for your in-depth comments which shaped my study to reach the current product. Prof. Yasir, thank you for providing me the Doctoral training opportunity.

I am grateful to the Netherlands Fellowship Programme (NFP) for funding this research. I would like to thank Jolanda Boots from IHE Delft for her support in all administrative work. I record my deep gratitude to Silvia for supporting me when I was ill.

I consider myself fortunate indeed to have the opportunity to pursue my study at IHE, an institute with a multi-cultural environment. I had a chance to meet good and inspiring people from all over the world, to exchange knowledge and expand my network. I thank friends from the IHE PhD group, Chol Abel, Mawiti Infantri Yekti, Yasir Salih, Zahra Naankwat Musa, and Mario Castro Gama, for sharing thoughts and experience. My deep sense of thanks to my good friends and accommodation mates, Eiman Fadol, Jakia Akter and Marmar Badr, I was blessed by your accompany during my stay in Delft. I cannot forget the blessed accompany of Shaza Jameel, Salman Adam, Omer Musa, and Sara Altayeb.

My thanks extend to the Sudanese Community in Netherlands in general and Delft, Ghada, Tayseer, and Eng. ALfatih family, for providing sense and warmth of home. Special thanks to Mr. Fikri Kurror, the first person I met in Netherlands.

This acknowledgement would not be completed without mentioning my friends from Sudan, Nazik, Nayla and Zeinab, I owe deep sense of gratitude to your accompany and encouragement which helped me a lot to continue my journey. Nayla, thank you for sharing your thoughts and rich experience. Nazik, I appreciate your care. Zeinab, I have been gifted by meeting you for the first time at IHE and having your rock steady support.

I owe my deepest gratitude to my family, both extended and small. My parents, without your encouragements and unconditional support I would never come to what I have done. There are no proper words to convey and express my gratitude for your wisdom. My brothers, Mohamed, Mazin and Mutaz, thank you for all what you did and I cannot even tell. Mazin, can't forget your effort to facilitate model computations in coputers of limited capacities. My small family, spouse Masoud and children, Lana, Ahmed, Mohamed and Momin, I am immeasurably grateful for your patience and sacrificing when I was away from home.

I cannot end my acknowledgement without thanking the soul of my uncle Hassan Shalabi Mukhtar, a teacher who fought for Nubian's female education. His love and encouragement for education is behind the success of many women in Sudan.

SUMMARY

The Eastern Nile (EN) riparian countries Egypt, Ethiopia and Sudan are currently developing several reservoir projects to contribute to the needs for energy and food production in the region. The Nile Basin, particularly the Eastern Nile Sub-basin, is considered one of the international river systems with potential conflicts between riparian countries. Yet, the Eastern Nile is characterized by the high dependency of downstream countries on river water generated in upstream countries.

In the absence of formal mechanisms for collaboration, the transboundary nature of the EN basin makes sound water resources development very challenging. The large seasonal and inter-annual variability of the river flow exacerbate those challenges. A further complication is the high sediment load in the EN Rivers, particularly during the high flow season. The operation of most of the reservoirs have been developed without sufficiently considering sediment management.

The Nile basin water resources have been extensively studied during the last 100 years or more, for planning and management purposes, in particular with regard to the use of irrigation water in the downstream part of the basin, though recently some studies have also focused on use of water for hydropower generation in the upper parts. These studies show that there is no convergence of development plans emerging among the Nile riparian countries. Another challenge is that the current reservoir optimization and simulation models cannot handle the temporal and spatial variations and implications of sediment deposition of multiple multi-purpose reservoirs.

The aim of this PhD research is to analyse the long-term impacts of water resources development on water quantity and reservoir sedimentation, considering different system management options and operating rules of existing dams. To identify knowledge gaps regarding modelling of Nile water resources, the first part of this PhD research reviewed water resource models applied in the Nile Basin, distinguishing between simulation, optimization and combined simulation and optimization models. The review shows that the political dimensions and societal, economic and environmental risks associated with water resources development have not been fully addressed in the Nile basin models, which could possibly explain why certain developments are opposed by some riparian countries. The output of this part was important to guide future research on water resources planning and management in the Nile.

The second part of the PhD study investigated the implications of water resources development on water availability in the Eastern Nile basin, for hydropower generation and irrigation water demands. The implications were assessed both at country and regional levels, using scenario analysis within a river basin simulation model. Twelve scenarios were investigated including: new dam developments; new irrigation schemes;

and different options for dam operation, i.e. unilateral versus cooperative transboundary management of dams. A RIBASIM model of the Eastern Nile was built that included twenty dams and twenty-one irrigation schemes, and used historical data of the hydrology of 103 years at a monthly time step as input. The operating rules of existing dams were assumed to remain unchanged. Four indicators were used for evaluating the performance of the system: hydropower generation [MWh/yr], reliability of irrigation supply [%], reservoir net evaporation [10^6 m^3/yr] and flow regimes of rivers [m^3/s].

The third part of the PhD study aimed to analyse the optimal operation scenarios for water resources management in the EN to satisfy hydropower generation and irrigation requirements. A hydro-economic optimization model based on Genetic Algorithm and a deterministic optimization approach was developed and used to determine the maximum benefits for two scenarios: (i) non-cooperative management of dams in the EN basin by the riparian countries, and (ii) cooperative management of those dams among the riparian countries. The EN system was optimized in the cooperative management scenario as one system and generates system-wide economic returns. In the non-cooperative management scenario, the system within each country was optimized separately, releases from the optimal system state in the upstream country were used as regulated inflows for optimizing the downstream country's system. The simulation results of current operation of the existing system were used as base scenario to compare the results of optimization. The hydro-economic model covered all currently (2020) existing hydraulic infrastructures in the EN (TK5, Roseires, Sennar, J. Aulia, Settit, K. Girba, Merowe and Aswan High Dam) and the existing irrigation schemes plus those attached to the Settit dam in Sudan (168,000ha). Subsequently, the Grand Ethiopian Renaissance Dam (GERD) was included in the optimization as an alternative scenario. The Eastern Nile system in Sudan was assumed to be constrained by the 1959 Agreement in all scenarios, which limits water withdrawals in Sudan to 18.5×10^9 m^3/yr.

The simulation results show that, managing the existing EN system in a cooperative transboundary manner without changing the operating rules of existing reservoirs and without new irrigation development projects, the GERD would increase the hydropower generation in Ethiopia and Sudan by +1500% and +17%, respectively, and slightly reduce the hydropower generation in Egypt by -1% (long term average values). The model runs show that unilateral management of the existing system following the installation of the GERD would not affect the hydropower generation significantly compared to cooperative management because the GERD would be operated for hydropower generation only, which is largely a non-consumptive water use.

The results of optimizing the operation rules of the EN system, assuming cooperative management of the existing system, show that hydropower generation can be increased in Ethiopia and Sudan by 1100% and 25%, respectively, following the construction of GERD, compared to the base case. In contrast to the simulation results, the optimization results show an increase of hydropower generation in Egypt (+8%) when GERD gets operational and the whole system is cooperatively managed, compared to the base case. The optimised operation of the EN system with GERD results in a hydropower increase in Egypt and Sudan and a decrease in Ethiopia compared to the simulation results that assume the current operation of the existing system unchanged. This result can be explained by the relatively high economic return of hydropower generation as assumed in the model set-up, the large hydropower generation capacity of Aswan High Dam and its location at the downstream end of the system. Optimization results also show that unilateral system management would negatively impact the hydropower generation of Egypt compared to the base case (-3.5%) and compared to the optimization results for cooperative management (-11%), without a significant increase of hydropower generation for Ethiopia compared to the base case (1215%) and the optimization results for cooperative management (+2%). For Sudan, the results show that hydropower generation benefits from the presence of GERD in both management scenarios. Non-cooperative management of the system, along with the internal trade-off between irrigation and hydropower, would negatively impact irrigation supply in Sudan. The internal trade-off in Sudan is attributed to the location of irrigation demand upstream of Merowe dam, the largest hydropower generation dam in Sudan (1,250 MW). The results also show that the supply reliability of existing and planned irrigation schemes in Sudan would practically not be influenced by the GERD, but would reduce to 92% when upstream dam developments and new irrigation expansion materialize in Ethiopia. Similarly, the existing irrigation schemes in Egypt would experience a deficit of 9% in the supply following upstream irrigation expansion. Unilateral management of a fully developed basin would increase the rate of evaporation losses in the basin by +15%, compared to cooperative management. Full development of the EN basin refers here to the proposed hydro dams on the Main Nile in Sudan (Dal, Sheriq, Kajabar and Sbloga dams) and the Blue Nile in Ethiopia (GERD, Mendaya, Beko Abo and Karadobi dams), and irrigation schemes in both countries. In general, water resources developments would have considerable but varying impacts on the countries in the long-term. Further impacts would be expected during the filling stage depending on the filling procedure of the GERD; however, assessing the filling stage was beyond of the scope of this study.

The fourth and last part of the PhD study focused on developing a new model for a multi-objective multiple reservoir system optimization and simulation that includes sediment management. The model constitutes three modules; optimization, reservoir operation and sediment management simulation modules. The trap efficiency concept was applied for sediment simulation. Optimization was based on the Genetic Algorithm available in the optimization tool box of MATLAB. All modules were coded in MATLAB 2015-b. The model was applied to optimize the operation of Roseires Reservoir in the Blue Nile River in Sudan (single reservoir system). The operation of Roseires reservoir was optimized for three objective functions: maximizing the economic return from hydropower generation, maximizing the release of water for irrigation, and minimizing sediment deposition through sluicing. Four scenarios were compared to assess the benefits from optimizing the operation: (i) the current operation policy, (ii) maximizing hydropower and irrigation benefits, (iii) optimising sediment management, and (iv) a weighted function to support achieving all three objectives collectively. The results show that the combined economic return of hydropower and irrigation increases by 5% over 20 years when sediment management is considered in reservoir operation, both as sole objective or with other water uses, compared to ignoring the sediment management component. When sediment management is not included, the storage capacity of the reservoir would be halved in 20 years and irrigation water deficits would occur 10 % of the time (during 20 years). The results also show that, compared to the existing operation practice, which favours sediment management during the flood season, sediment deposition could be further reduced, which would benefit irrigation and hydropower production in the long run. Trap efficiency could reach 25% compared to 39.5% of the existing practice.

This study contributed to fill relevant knowledge gaps through a better understanding of the methods needed for a complex system of multipurpose reservoirs, considering both water quantity and sediment load. More specifically, the developed models for water management allowed assessing the applicability of a combined optimization and simulation approach for a real complex system including reservoir sedimentation problem. This study thus contributes to closing the gap between real-world cases and pure research problems.

The study also comparatively quantified the impacts of water resources development in the EN basin and assisted in identifying system management options at different levels (regional and country level). As a result, it is shown that developing a collaborative and unified perspective of the countries towards new projects can be beneficial for all. In addition, the study proposed new operation rules for improving operation of the current system when new infrastructures are developed and operated either unilaterally or cooperatively. Distribution of the benefits between countries were quantified for both cooperative and non-cooperative management options. Evidence based policies are the basis for sustainable development and peace in the region, and this study attempted to provide a basis for this.

The findings indicated that the optimal operation of the system for hydropower generation and irrigation following infrastructure development would shift towards hydropower generation, unlike the current operation, where irrigation is the dominant objective. This shift resulted from many interrelated aspects that need to be explored more in future studies, such as the largely non-consumptive nature of hydropower and its relatively high economic return, as well as the specific locations of hydropower dams in the basin. The location and objectives of proposed dams would need further studies for basin-wide better use of available water and collective benefits. Cropping patterns of irrigation projects and water management at field level should be included in future reservoir operation studies. Future studies should also include the Main Nile system downstream of Aswan High Dam.

Future research along these lines should be continued to include advanced sediment transport models for sediment management simulations in the EN multi-reservoir system management. Simple trap efficiency models can be used for planned dams that do not have observed data, while sediment transport models can be calibrated and more accurately estimate the trap efficiency for existing reservoirs. The operation of reservoirs can be optimised further when sediment management is included.

SAMENVATTING

Egypte, Ethiopië en Soedan, oeverstaten van de Oostelijke Nijl (ON) rivier, ontwikkelen momenteel verschillende reservoirprojecten om bij te dragen aan de behoeften aan energie- en voedselproductie in de regio. De Nijl rivier, en met name het stroomgebied van de Oostelijke Nijl, wordt beschouwd als een van de grensoverschrijdende rivieren met potentiële conflicten tussen oeverstaten. Toch wordt de Oostelijke Nijl gekenmerkt door benedenstroomse landen die afhankelijk zijn van rivierwater dat zijn oorsprong heeft in bovenstroomse landen.

Bij gebrek aan formele samenwerkingsmechanismen, maakt het grensoverschrijdende karakter van het ON-bekken een deugdelijke water ontwikkeling zeer uitdagend. De grote seizoensgebonden zowel als jaarlijkse variabiliteit van de rivierafvoer maakt dit nog ingewikkelder. Een andere complicatie is de hoge sedimentlast in de ON-rivieren, vooral tijdens het seizoen met hoge afvoeren. Het beheer van de meeste reservoirs houdt nog onvoldoende rekening met deze sediment problematiek.

De water situatie van het stroomgebied van de Nijl is de afgelopen 100 jaar of langer uitgebreid bestudeerd voor planning- en beheerdoeleinden, met name met betrekking tot het gebruik van irrigatiewater in het benedenstroomse deel van het bekken, hoewel recentelijk enkele onderzoeken ook gericht waren op het gebruik van water voor de opwekking van waterkracht in de bovenstroomse landen. Deze studies tonen aan dat er geen convergentie is van ontwikkelingsplannen tussen de oeverstaten van de Nijl. Een andere uitdaging is dat de huidige optimalisatie- en simulatiemodellen voor reservoir-beheer de temporele en ruimtelijke variaties en implicaties van sedimentafzetting van meerdere multifunctionele reservoirs niet aankunnen.

Het doel van dit doctoraatsonderzoek is het analyseren van de langetermijneffecten van water ontwikkeling op het beschikbare water en de sedimentatie van reservoirs, rekening houdend met verschillende opties voor systeembeheer en beheer-regels van bestaande dammen. Om lacunes in de kennis met betrekking tot het modelleren van water in de Nijl te identificeren, beoordeelde het eerste deel van dit proefschrift de water modellen die op het stroomgebied van de Nijl zijn toegepast, waarbij onderscheid wordt gemaakt tussen simulatie, optimalisatie en gecombineerde simulatie- en optimalisatiemodellen. Uit de evaluatie blijkt dat de politieke dimensies en de maatschappelijke, economische en milieurisico's die samenhangen met de water ontwikkelingen niet volledig zijn geadresseerd in modellen van de Nijl rivier, wat mogelijk zou kunnen verklaren waarom bepaalde ontwikkelingen worden tegengewerkt door sommige oeverstaten. Dit deel van het proefschrift was belangrijk als leidraad voor toekomstig onderzoek naar de planning en het beheer van water in de Nijl.

Het tweede deel van het proefschrift onderzocht de implicaties van de water ontwikkelingen op de beschikbaarheid van water in het Oostelijke Nijlbekken, voor de opwekking van waterkracht en de vraag naar irrigatiewater. De implicaties werden beoordeeld op zowel nationaal als regionaal niveau, met behulp van scenario-analyse binnen een stroomgebied simulatiemodel. Twaalf scenario's zijn onderzocht, waaronder nieuwe damontwikkelingen, nieuwe irrigatieprojecten, en verschillende opties voor dambeheer, namelijk eenzijdig versus coöperatief grensoverschrijdend beheer van dammen. Een RIBASIM-model van de Oostelijke Nijl is opgezet dat twintig dammen en eenentwintig irrigatieprojecten omvat, historische gegevens van de hydrologie van 103 jaar gebruikt, en dat een maandelijkse tijdsstap heeft. Aangenomen werd dat de regels van het beheer van bestaande dammen ongewijzigd bleef. Er zijn vier indicatoren gebruikt om de resultaten van het systeem te evalueren: waterkrachtopwekking [MWh/jaar], betrouwbaarheid van de irrigatievoorziening [%], netto-verdamping van reservoir water [10^6 m^3/jaar] en afvoer regimes van rivieren [m^3/s].

Het derde deel van het proefschrift was gericht op het analyseren van de optimale scenario's voor het water beheer in de ON om te voldoen aan de vereisten voor waterkrachtopwekking en irrigatie. Een hydrologisch-economisch optimalisatiemodel was ontwikkeld op basis van genetisch algoritme en een deterministische optimalisatiebenadering. Dit optimalisatie model is gebruikt om het maximale profijt voor twee scenario's te bepalen: (i) niet-coöperatief beheer van dammen in het ON-bekken door de oeverstaten, en (ii) coöperatief beheer van die dammen tussen de oeverstaten. Het ON-systeem is in het scenario voor coöperatief beheer geoptimaliseerd als één systeem en genereert systeem-brede economische rendementen. In het niet-coöperatieve beheerscenario werd het systeem binnen elk land afzonderlijk geoptimaliseerd. De rivierafvoeren resulterend van de optimale systeemstatus in het bovenstroomse land werden gebruikt als gereguleerde instromen om het systeem van het naastgelegen benedenstroomse land te optimaliseren. De simulatieresultaten van het huidige waterbeheer van het bestaande systeem werden gebruikt als basisscenario waarmee de resultaten van de optimalisatie vergeleken werden. Het hydrologisch-economische model omvatte alle momenteel (2020) bestaande reservoirs in de ON (TK5, Roseires, Sennar, J. Aulia, Settit, K. Girba, Merowe en Aswan High Dam) en de bestaande irrigatieprojecten alsmede dat verbonden is aan de Settit dam in Sudan (168.000ha). Vervolgens werd de Grand Ethiopian Renaissance Dam (GERD) als alternatief scenario meegenomen in de optimalisatie. Het Oostelijke Nijl systeem in Soedan werd in alle scenario's beperkt door de Overeenkomst van 1959, die de wateraafvoer in Soedan beperkt tot $18,5 \times 10^9$ m^3/jaar.

De simulatieresultaten tonen aan dat de GERD, door het bestaande ON-systeem op een coöperatieve manier grensoverschrijdend te beheren, de opwekking van waterkracht in Ethiopië en Soedan met + 1500% en + 17% zou verhogen en de opwekking van waterkracht in Egypte met -1% zou verminderen (gemiddelde lange-termijnwaarden). In dit scenario bleven de regels van het beheer van de bestaande reservoirs ongewijzigd, en werden geen nieuwe irrigatieprojecten meegenomen. De modelresultaten tonen aan dat eenzijdig beheer van het bestaande systeem na de bouw van de GERD de opwekking van waterkracht niet significant zou beïnvloeden in vergelijking met coöperatief beheer, omdat de GERD alleen zou worden gebruikt voor de opwekking van waterkracht, wat grotendeels een niet-consumptief watergebruik is.

De resultaten van het optimaliseren van de beheer regels van het ON-systeem, uitgaande van coöperatief beheer van het bestaande systeem, tonen aan dat de opwekking van waterkracht in Ethiopië en Soedan kan worden verhoogd met respectievelijk 1100% en 25%, na de bouw van GERD, vergeleken met de basisscenario. In tegenstelling tot de simulatieresultaten, laten de optimalisatieresultaten een toename zien van de opwekking van waterkracht in Egypte (+8%) wanneer GERD operationeel wordt en het hele systeem coöperatief wordt beheerd, vergeleken met het basisscenario. De geoptimaliseerde werking van het ON-systeem met GERD resulteert in een toename van waterkracht in Egypte en Soedan en een afname in Ethiopië in vergelijking met de simulatieresultaten waarbij de huidige werking van het bestaande systeem onveranderd bleef. Dit resultaat kan worden verklaard door het relatief hoge economische rendement van de waterkrachtopwekking zoals verondersteld in het optimalisatiemodel, de grote waterkrachtopwekkingscapaciteit van Aswan High Dam en de benedenstroomse locatie van deze dam in het systeem. Optimalisatieresultaten tonen ook aan dat eenzijdig systeembeheer een negatieve invloed zou hebben op de waterkrachtproductie van Egypte in vergelijking met het basisscenario (-3,5%) en vergeleken met de optimalisatieresultaten voor coöperatief beheer (-11%), zonder een significante toename van de waterkrachtproductie voor Ethiopië vergeleken met het basisscenario (1215%) en de optimalisatieresultaten voor coöperatief beheer (+2%). Voor Sudan laten de resultaten zien dat de opwekking van waterkracht profiteert van de aanwezigheid van GERD in beide beheerscenario's. Niet-coöperatief beheer van het systeem, samen met de interne trade-off tussen irrigatie en waterkracht, zou een negatief effect hebben op de irrigatievoorziening in Sudan. De interne trade-off in Sudan wordt toegeschreven aan de locatie van de irrigatievraag stroomopwaarts van de Merowe-dam, de grootste waterkrachtcentraledam in Sudan (1.250 MW). De resultaten tonen ook aan dat de leveringsbetrouwbaarheid van bestaande en geplande irrigatieprojecten in Sudan praktisch niet wordt beïnvloed door de GERD, maar zou dalen tot 92% wanneer damontwikkelingen en nieuwe irrigatie-expansie in bovenstrooms Ethiopië plaatsvinden. Evenzo zouden de bestaande irrigatieprojecten in Egypte een tekort van 9% in het aanbod ondervinden als gevolg van de uitbreiding van de irrigatie stroomopwaarts. Eenzijdig beheer van een volledig ontwikkeld bekken zou het percentage verdampingsverliezen met +15% verhogen in vergelijking met coöperatief beheer. De volledige ontwikkeling van het ON-bekken verwijst hier naar de voorgestelde dammen op de Main Nile in Sudan

(Dal, Sheriq, Kajabar en Sbloga) en de Blauwe Nijl in Ethiopië (GERD-, Mendaya, Beko Abo en Karadobi) en irrigatieprojecten in beide landen. In het algemeen zouden deze water ontwikkelingen op de lange termijn aanzienlijke maar wisselende gevolgen hebben voor de landen. Verdere effecten zijn te verwachten tijdens de fase van het vollopen van de GERD, maar dat is afhankelijk van de vulprocedure van de GERD; dit viel echter buiten het bestek van deze studie.

Het vierde en laatste deel van het proefschrift was gericht op het ontwikkelen van een nieuw model voor een meerdoelige optimalisatie en simulatie van meerdere reservoirsystemen met sedimentbeheer. Het model bestaat uit drie modules; modules voor optimalisatie, reservoirbeheer en simulatie van sedimentbeheer. Het concept van trapefficiëntie werd toegepast voor sediment-simulatie. Optimalisatie was gebaseerd op het genetische algoritme dat beschikbaar is in de optimalisatie toolbox van MATLAB. Alle modules zijn gecodeerd in MATLAB 2015-b. Het model is toegepast om de werking van de Roseires dam in de Blauwe Nijl river in Sudan (een systeem met één reservoir) te optimaliseren. De werking van het Roseires-reservoir is geoptimaliseerd voor drie doelen: het maximaliseren van het economische rendement van de opwekking van waterkracht, het maximaliseren van water voor irrigatie en het minimaliseren van sedimentafzetting door water weg te sluizen. Vier scenario's werden vergeleken om de voordelen van het optimaliseren van de operatie te beoordelen: (i) het huidige exploitatiebeleid, (ii) het maximaliseren van de voordelen van waterkracht en irrigatie, (iii) het optimaliseren van sedimentbeheer, en (iv) een gewogen functie ter ondersteuning om alle drie de doelstellingen tegelijk te bereiken. De resultaten laten zien dat het gecombineerde economische rendement van waterkracht en irrigatie over 20 jaar met 5% toeneemt wanneer sedimentbeheer wordt meegenomen in het beheer van het reservoir, zowel als het als enig doel wordt gesteld als dat het met andere watergebruiken wordt gecombineerd, in vergelijking met het negeren van de component voor sedimentbeheer. Als sedimentbeheer niet is inbegrepen, zou de opslagcapaciteit van het reservoir in 20 jaar worden gehalveerd en zou 10% van de tijd (gedurende 20 jaar) een tekort aan irrigatiewater optreden. De resultaten tonen ook aan dat, in vergelijking met de bestaande beheerpraktijk met actief sedimentbeheer tijdens het seizoen met hoge afvoeren, de sedimentafzetting verder zou kunnen worden verminderd, wat de irrigatie en de productie van waterkracht op lange termijn ten goede zou komen. De trapefficiëntie zou 25% kunnen bereiken vergeleken met 39,5% in de bestaande praktijk.

Deze studie heeft bijgedragen aan het opvullen van relevante kennislacunes door een beter begrip van de methoden die nodig zijn voor een complex systeem van multifunctionele reservoirs, rekening houdend met zowel de waterhoeveelheid als de sedimentlast. Meer specifiek maakten de ontwikkelde modellen voor waterbeheer het mogelijk om de toepasbaarheid van een gecombineerde optimalisatie- en simulatiebenadering te beoordelen voor een bestaand complex systeem inclusief reservoirsedimentatie. Deze studie draagt dus bij aan het dichten van de kloof tussen praktijkgevallen en pure onderzoeksproblemen.

De studie kwantificeerde ook de effecten van de water ontwikkelingen in het ON-bekken, op een vergelijkende manier, en hielp bij het identificeren van opties voor systeembeheer op verschillende niveaus (regionaal en landelijk). De bevindingen tonen aan dat de ontwikkeling van een gezamenlijk en verenigd perspectief van de landen voor nieuwe projecten voor elk voordelig kan zijn. Daarnaast stelde de studie nieuwe beheer-regels voor om de werking van het huidige systeem te verbeteren wanneer nieuwe waterwerken worden ontwikkeld en geëxploiteerd, hetzij eenzijdig of coöperatief. De verdeling van de voordelen over de landen werd gekwantificeerd voor zowel coöperatieve als niet-coöperatieve beheeropties. Beleid gebaseerd op wetenschappelijk bewijs vormt de basis voor duurzame ontwikkeling en vrede in de regio, en met deze studie is getracht hiervoor een fundering te leggen.

De bevindingen gaven aan dat het optimale beheer van het systeem voor de opwekking van waterkracht en irrigatie na de ontwikkeling van nieuwe waterwerken zou verschuiven naar de opwekking van waterkracht, in tegenstelling tot de huidige beheerpraktijk, waarbij irrigatie het overheersende doel is. Deze verschuiving kan verklaard worden door een samenspel van aspecten die in toekomstige studies nader moeten worden onderzocht, zoals het grotendeels niet-consumptieve karakter van waterkracht en het relatief hoge economische rendement daarvan, evenals de specifieke locaties van waterkrachtdammen in het bekken. De locatie en doelstellingen van de voorgestelde dammen zouden verder moeten worden onderzocht om het beschikbare water in het hele stroomgebied beter te benutten zowel als de collectieve voordelen. Gewaspatronen van irrigatieprojecten en waterbeheer op veldniveau moeten worden opgenomen in toekomstige studies over het gebruik van reservoirs. Toekomstige studies zouden ook het Main Nile-systeem benedenstrooms van de Aswan High Dam moeten omvatten.

Toekomstig onderzoek langs deze lijnen zou ook sedimentbeheer in het ON-systeem met meerdere reservoirs moeten omvatten, gebruikmakend van geavanceerde sedimenttransportmodellen. Eenvoudige trapefficiëntiemodellen kunnen worden gebruikt voor geplande dammen waarvoor nog geen empirische gegevens beschikbaar zijn, terwijl voor bestaande reservoirs sedimenttransportmodellen kunnen worden gekalibreerd om de trapefficiëntie nauwkeuriger te kunnen schatten. Het beheer van reservoirs kan verder worden geoptimaliseerd wanneer sedimentbeheer is inbegrepen.

CONTENTS

1

INTRODUCTION

1

1.1 Background

Rivers are multi-dimensional systems, including physical, ecological and economic systems. They are politically significant when they are shared between nations (Sadoff & Grey, 2002). Water allocation in trans-boundary river basins is a critical and complex issue when water is scarce (Asfaw & Saiedi, 2011; Barrow, 1998; Dinar et al., 2007). The complexity is characterised by conflicting objectives within and between riparian states (Rani & Moreira, 2010), adding to the inherent uncertainty of stream flows and demands, and the interdisciplinary nature of addressing water management issues. This is particularly true in the case of the Eastern Nile River basin; a sub-basin of the Nile river basin, one of the largest and least developed trans-boundary river basins in the world.

The Eastern Nile basin is a trans-boundary basin shared by four countries: Ethiopia, South Sudan, Sudan and Egypt and covers approximately more than one half of the Nile basin. The Eastern Nile (EN) basin is the source of more than 80% of the Nile river flow. The basin is characterized by many trans-boundary issues that urge the needs for water resources development and at the same time challenge water resources management. The countries of the EN basin are characterized by rapid population growth, widespread poverty and political instability. Water management in the basin is challenged by competing water uses among sectors, and among riparian states, as well as often low efficiencies of water use exacerbated by increasing environmental degradation. The rivers of the basin are characterized by high temporal and spatial flow variability. Climatic variability and uncertainty with respect to future climate change poses serious challenges towards water resources management (A.P. Georgakakos, 2007; Goor et al., 2010; Griensven et al., 2012; Ribbe & Ahmed, 2006; Sayed, 2008). High sediment loads, a dimension neglected in most studies, and the scarcity of data and lack of data sharing protocols add to the challenges of sound water resources development. The increased demand for water, combined with ambitious economic growth policies in Eastern Nile riparian countries, have resulted in a myriad of, largely un-coordinated, water resources developments and plans.

The basins encountered a drastic environmental degradation represented by deforestation and high erosion leading to the loss of upstream land, increased flood risk and sediment load which in turn affects the downstream infrastructures (i.e. reservoir sedimentation) and irrigation schemes (i.e. clogging of irrigation canals and reducing the agricultural productivity) (Dinar & Nigatu, 2013; Schleiss et al., 2016). For instance, Roseires, Sennar and Khashm Elgirba dams in Sudan (downstream state) have lost about 60 %, 34 % and 43 % of their storage capacity, respectively (ENTRO, 2007; Gismalla, 2009). Sediments have also created difficulties for the management of the Gezira irrigation scheme (Osman, 2015).

However, the Eastern Nile is endowed with huge hydropower and food production potentials that can be generated from cooperative water resources development and management. Upstream countries possess potential of hydropower generation, while the downstream ones are blessed with ample irrigable fertile soil. Only 3% of the basin's hydro-electricity potential has been developed so far (Habteyes et al., 2015). Water resources development for hydropower generation and irrigated agriculture needs cooperation between riparian countries because of the limited water availability. Full cooperation in the EN basin is however not practiced yet (S. M. A. Salman, 2016).

Cooperative and non-cooperative management of trans-boundary river basins have been debated by scholars for many years (Dinar & Nigatu, 2013). Cooperation is shown to produce significant benefits compared to non-cooperation (Dinar & Nigatu, 2013; Dombrowsky, 2009b). However, riparian states tend to move towards non-cooperation as the scale of benefits may not justify the cost of cooperation (Wu & Whittington, 2006). Sadoff and Grey (2002) categorized the benefits that could yield from cooperation into four groups: benefits to the river resulting from better management of ecosystems, benefits from the rivers resulting in increased energy and food production, benefits from a reduction of the costs because of rivers resulting from improved cooperation between riparian states, and benefits from cooperation beyond the river resulting from the economic integration between states.

The Nile Basin, and in particular the Eastern Nile Sub-basin, is considered as one of the international river systems with potential water conflicts between riparian countries (Samaan, 2014; Wu & Whittington, 2006). In common with other international rivers, current tensions in the Eastern Nile Sub-basin and the whole Nile Basin are triggered by water availability that is insufficient to satisfy the water needs of all planned development projects. Each of the basin countries is unilaterally developing water resources projects to meet the increasing demand for energy and economic growth (Goor, et al., 2010; Jeuland, 2010; Whittington et al., 2005). However, unilateral management limits the potential benefits from transboundary water resources, which can be extended beyond shared water system management (Cascão, 2009; Matthew P. McCartney & Menker Girma, 2012). The unique feature of the tensions in the Eastern Nile Basin is that downstream countries have a high dependency on the water generated in upstream countries (Wu & Whittington, 2006).

In the absence of formal mechanisms for collaboration in the basin, the impacts of unilateral management on each state need to be quantified and thereafter cooperative management can be introduced as best alternative to provide win-win situations among the states. Assessing water-related technical, socio-economic issues in the basin is complex, and therefore requires specialized river basin modelling tools (Belachew et al., 2015).

3

Nile basin water resources development and management has been studied extensively for more than one century. Sir William Wilcocks in 1890 promoted basin wide demand coordination, in an attempt to prepare the Nile regulation plan (Barrow, 1998). A British plan known as "Century Storage Scheme" for full Nile water resources development was published in 1920 (Wolf & Newton, 2013). Most of the available studies are based on control infrastructures proposed in both the Nile Valley Plan study and the United States Bureau of Reclamation (USBR) study conducted in 1958 and 1964, respectively. The results of these studies have not found consensus among the Nile basin parties owing to many reasons. Among these reasons are the inconsistent, fragmented knowledge of the basin and limitation of data and information sharing (Matthew P. McCartney & Menker Girma, 2012).

Several modelling studies of the Nile have been conducted to support decision making of transboundary water management (Arjoon et al., 2014; P. Block & Strzepek, 2010; P. J. S. Block, Kenneth Rajagopalan, Balaji, 2007; A.P. Georgakakos, 2007; Goor, et al., 2010; Guariso et al., 1981; Guariso & Whittington, 1987; Habteyes, et al., 2015; Jeuland et al., 2017; Y. Lee et al., 2012; Satti et al., 2014; Whittington, et al., 2005), but very few (Abdallah & Stamm, 2013; Ali, 2014; Yoon Lee et al., 2012; Yasir Abbas Mohamed, 1990) have considered the effect of reservoir sedimentation in the water resources development plans. Although good insights of the system and expected impacts of developments have been gained, still the picture is not fully understood for different topologies and probabilities of (future) river flows. Therefore, studying water resources development options in a regional context is still important to quantify the impacts both at regional and at country level. Limited use of appropriate analytical tools as a result of limitations of the financial, institutional and human capacity, which is a common problem throughout Africa, might also be a reason (Matthew P McCartney, 2007).

1.2 OBJECTIVES

The main objective of this PhD study is to analyse the long-term impacts of water resources development on water quantity and reservoir sedimentation of the EN, considering different system management options. The specific objectives are:

1) To identify an appropriate modeling approach of the reservoir system in the Eastern Nile basin.
2) To assess the implication of new dam constructions in the Eastern Nile for water availability for hydropower and irrigation at national and regional levels.
3) To develop optimal operation rules for the multi-purpose multi-reservoir system of the Eastern Nile basin with and without consideration of reservoir sedimentation.

1.3 THESIS OUTLINES

The thesis includes seven chapters. Chapter two describes the Eastern Nile basin. The main sub-basins and their topographic, climatic and hydrologic conditions are outlined. Description of the main infrastructures and irrigation projects as well as the cooperative programmes and projects for water resources development are provided.

Chapter three presents a literature review of the application of river basin modelling to support Nile basin water management.

Chapter four evaluates different options of water resources development considering different levels of cooperative management using a river basin simulation model and scenario analyses. The impacts of water resources development on hydropower generation, irrigation supply, reservoir evaporation and transboundary inflows are investigated.

Chapter five assesses the optimal operation of the Eastern Nile basin system after the GERD development at country and basin-wide levels using Genetic Algorithm. The optimization focuses on maximizing hydropower generation and irrigation supply.

Chapter six investigates the optimization of the operation of the Eastern Nile system including sediment management. The development of a new modelling approach is described, which is applied to Roseires dam on the Blue Nile in Sudan.

Finally, chapter seven summarises the main findings and conclusions.

2
STUDY AREA

The study area is a major part of the Nile river basin. As many issues in the study area are applicable to the entire Nile basin, this chapter starts with a brief introduction of the Nile River Basin.

2.1 THE NILE RIVER BASIN

The Nile River is the longest river in the world, extending about 6700 km from the source, headwaters in eastern Africa at more than 4000 m.a.s.l. (meters above sea level) at the headwaters to the sea level at the Nile Delta in Egypt (NBI, 2012). It flows through eleven riparian countries (**Figure 2.1**): Burundi, Democratic Republic of Congo, Kenya, Rwanda, Tanzania, Uganda, South Sudan, Eritrea, Ethiopia, Sudan, and Egypt, and is home to more than 300 million people (Sayed, 2008). The average annual natural flow is 84×10^9 m3/yr as measured at Aswan High Dam, with $1,700 \times 10^9$ m3/yr of rainfall (Ribbe & Ahmed, 2006; Sayed, 2008). The climate of the basin varies significantly; it encompasses five climate zones that vary from tropical, to subtropical, semi-arid, arid and Mediterranean zones. The river yields water from only 20% of its catchment area, because more than half of its course flows through semi-arid and arid areas with hardly or no effective rainfall.

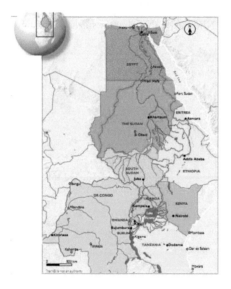

Figure 2.1 Location of the Nile River Basin(Source: NBI,(2012))

The two main sub-basins in the Nile basin are the Eastern Nile and the Nile Equatorial Lake. The Eastern Nile Basin with an area of about 1,657,845 km2 is the major sub-basin of the Nile, spanning four countries: South Sudan, Ethiopia, Sudan and Egypt (ENTRO, 2007). The main rivers of the basin are the Blue Nile, White Nile, and Main Nile, accumulating runoffs of four sub-basins: Blue Nile (56%), Atbara (15%), White Nile-Albert (14%) and Sobat (15%) as depicted in **Figure 2.2 (a)**.

2.2 THE EASTERN NILE SUB-BASINS

The research focuses on the Eastern Nile basin. An overview is provided on the most important features which dictate water resources availability and management of each sub-basin, including topography, climate, rainfall runoff and major water users.

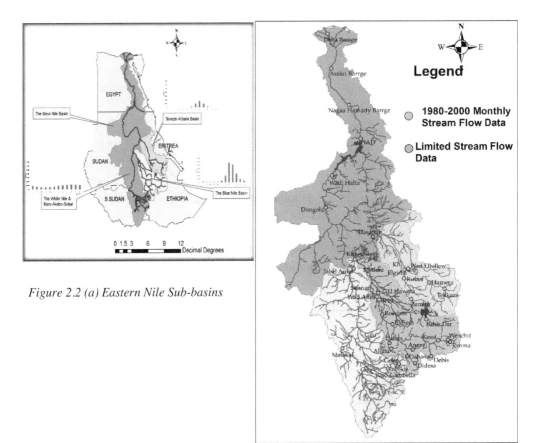

Figure 2.2 (a) Eastern Nile Sub-basins

(b) Stream flow data quality at different measuring stations

2.2.1 The Blue Nile sub-basin

The Blue Nile River originates from Lake Tana, in the Ethiopian highlands at an altitude of 1,830 m.a.s.l. It joins eight major tributaries, draining the south west and central Ethiopian highlands before it passes into Sudan. The total length of the Blue Nile course from Lake Tana to the Sudanese-Ethiopian border is 850 km, with a total drop in elevation of 1,300 m (Hassaballah, 2010), where it is steep in the plateau and flat at the border as shown in **Figure 2.3(a)**.

The climate of the Blue Nile river basin varies significantly between the headwaters in the highlands of Ethiopia and its confluence with the White Nile River at Khartoum in Sudan. The basin's highest rainfall is typically 2,000 mm/yr or more, but is characterized by high seasonality as well as annual variability. Moving northward through Sudan, rainfall gradually declines to about 200 mm/yr in Khartoum. The average potential evaporation rate varies from 1150 mm/yr at Lake Tana to 2500 mm/yr at Sennar region in south-east Sudan (Hassaballah, 2010). The average temperature fluctuates between 15-18°C in the highlands in Ethiopia, with variation and substantial increases northward in Sudan to reach 26.5°C.

The flow of the Blue Nile reflects the rainfall seasonality over the Ethiopian highlands. Two flow periods are apparent, the wet season and the dry season. The wet season or flood period is from July to October with peak flows in August and September. The dry season or low flow period extends from November to June. Due to the unimodal pattern of the rainfall in the basin, the annual Blue Nile hydrograph is characterised by a constant bell-shaped pattern, in spite of the annual flow volume variation as shown in **Figure 2.3(b)**. The average annual flow of the Blue Nile and its tributaries is 50 x10^9 m^3/year measured near the Ethiopia- Sudan border. The daily flow varies between 500x10^6 m^3/day in August and 10x10^6 m^3/day in April.

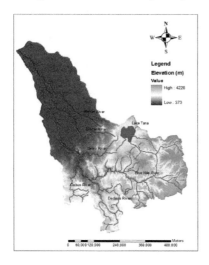
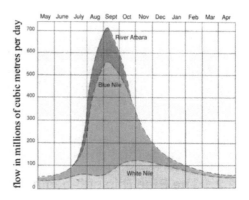

Figure 2.3: (a) The Blue Nile River and its   (b) The Nile River Hydrograph (Barron,
Tributaries   (2006))

There are seven flow measuring stations along the Blue Nile (**Figure 2.2(b)**). The Upper Blue Nile in Ethiopia has two main monitoring stations, namely at Bahir Dar and Kessi downstream Lake Tana with limited and incomplete records. In addition, there are eight gauges along the Blue Nile tributaries. In Sudan, five monitoring stations include Eldiem, Roseires, Sennar, Medani, and Khartoum. Roseires station has a substantial record length, while Eldiem Station at the Sudanese-Ethiopian border has a shorter series of records.

2.2.2 Baro-Akobo- Sobat sub-basin

Baro-Akobo-Sobat basin and lower part of the White Nile is located in the central part of the Nile basin (**Figure 2.4**). It covers an area of approximately $481,500 \text{ km}^2$ representing the catchment area of the Baro, Akobo, Pibor, Sobat, and lower White Nile up to the confluence with the Blue Nile at Khartoum. The drainage system of the basin includes rivers and large wetlands. The main river systems are Baro, Gila, Akobo, and Pibor. While Baro, Gila, Akobo originate from the Ethiopian Plateau, Pibor originates from South Sudan and northern Uganda. Large seasonal wetlands are formed by rivers spill.

The basin has a tropical climate with high rainfall in the mountainous area at elevations of 2,000 to 3,000 m.a.s.l. in Ethiopia, with declining rainfall northward to the flat plains in Sudan where the climate is arid. The wet season extends from May to October in the southern and eastern parts of the basin, rainfall being around 1,500 – 2,000 mm/yr. It decreases northward to start in July and end at September in the northern parts, with rainfall of about 150 mm/yr near Khartoum. The potential evapotranspiration follows a different trend, where it increases near Khartoum where the mean annual potential evaporation is recorded as 2920 mm/yr and decreases southward to reach 990 mm/yr (Yasir A.. Mohamed, 2011) (Shahin, 1985). The temperature exhibits a similar trend, with mean annual daily temperature range from 18°C at upper watershed to 30.5°C at Khartoum.

Figure 2.4 Location of Baro Akobo Sobat Basin (Source: ENTRO, (2007))

Half of the White Nile water is provided by the Baro Akobo Sobat. The White Nile reflects the seasonality of Baro Akobo Sobat, as the flow from Bahr El Jabel is rather steady. The major flow in the basin is supplied by the Baro River with an average annual flow of 9.5 x10^9 m^3/yr, while Pibor provides about 3.2 x10^6 m^3/yr (Yasir A.. Mohamed, 2011).

Few monitoring stations exist in the basin. Within Ethiopia, there are five hydrological stations (**Figure 2.2(b)**). In Sudan, there are several stations with short and incomplete records of flow. However, a discharge measurement series with sufficient length is available in Malakal (Yasir A.. Mohamed, 2011).

2.2.3 Tekeze - Atbara sub-basin

The Tekeze - Atbara basin (**Figure 2.5**) including three major tributaries originates from the central and north western highland plateaus of Ethiopia at an altitude above 3,000 m.a.s.l, declining to the low lands at less than 500 m.a.s.l. with flat and uniform topography at the confluence with the Main Nile in Sudan.

The Tekeze - Atbara basin encompasses four climate zones: moist sub humid, dry sub humid, semi-arid and arid climates identified from the highlands northward to the mouth in Sudan. The mean annual rainfall is about 1,000 mm/yr in the highlands in Ethiopia and decreases to less than 400 mm/yr at Elgirba station and 20 mm/yr at Atbara station. The mean annual temperature in the upper basin does not exceed 20°C, while at the confluence the temperature exceeds 30°C. Similar to the temperature trend, the mean annual potential evaporation in the highland plateau is below 2,000 mm/year and increases to reach 2,926 mm/yr in the low land area in Sudan (Sutcliffe & Parks, 1999) (Shahin, 1985). The mean annual flow at Atbara is 12×10^9 m^3/yr.

Figure 2.5 Location of Tekeze - Atbara Basin (Source: ENTRO, (2007))

Five gauging stations are available in the Tekeze - Atbara basin with 20 years data (1980-2000) (**Figure 2.2(b)**). In Ethiopia there are three stations, namely Humera, Embamadare and Zarima. In Sudan there are two river flow stations, i.e. Khashm Elgirba at Atbara River, and Kubur station at Upper Atbara River. There is flow gauge station in Wad Elhiliew at Settit (Tekeze) river with less than 20 years data.

2.2.4 The Main Nile sub-basin

The Main Nile sub-basin starts from the confluence of the Blue Nile and the White Nile at Khartoum at elevation of 400 m.a.s.l. to the Mediterranean Sea in Egypt (**Figure 2.6**). The sub-basin, occupying an area of 789,140 km^2, is characterized by a relatively flat topography. The main Nile River has only one tributary, namely Tekeze-Atbara River.

13

Figure 2.6 the Main Nile River Sub-basin (Source: ENTRO, (2007))

The climate ranges from arid climate at southern and central of the sub-basin to the Mediterranean Sea climate in the northern part of Egypt. The rainfall is negligible where about 65% of the sub-basin has an average annual rainfall of less than 50 mm/yr. The average annual rainfall varies from 200 mm/yr in Khartoum with rains occurring in autumn and decreases to 25 mm/yr in Cairo where it may rain in winter. The average annual rainfall starts to increase from Cairo to reach 200 mm/yr in Alexandria near the Mediterranean Sea. The average daily temperature varies from 30°C at Dongola and Aswan High Dam to 18°C in the coastal areas. Potential evaporation using Penman method is estimated at 2,924 mm/yr in Khartoum, 2,729 mm/yr at Dongola, 2,488mm/yr at Aswan High Dam and decreases to 1,800 mm/yr in Alexandria (Shahin, 1985; Sutcliffe & Parks, 1999).

The Main Nile average annual flow at Khartoum is 74.7×10^9 m^3/yr. At the confluence with the Atbara River the average annual flow increases to reach 86.7×10^9 m^3/yr. The flow decreases at Dongola to 85.5×10^9 m^3/yr due to losses of 1.2×10^9 m^3/yr between Hasnab and Dongola.

Four flow measuring stations are available in Sudan with a minimum 20 years data (from 1980) namely Tamaniat, Hasnab, Dongola, and Wadi Halfa (Figure 2.2(b)). In Egypt there are five gauge stations along the Main Nile with at least 20 years data. These gauges are at Aswan High Dam, Esna Barrages, Nagaa Hamady Barrage, Assiut Barrages and Delta Barrages. Many other gauge stations are available along the abstraction canals.

2.3 RESERVOIRS, HYDROPOWER PLANTS AND IRRIGATION PROJECTS

The Eastern Nile countries utilize the rivers mainly for irrigation, hydropower, domestic and industrial use, with irrigation having the largest consumptive water use demand. About 85% of the total Nile (blue) water consumption, estimated at $55.5 \times 10^9 \, m^3/yr$, is devoted to agriculture with an irrigated area of approximately 4.9×10^6 ha (Timmerman, 2005). About 97% of the irrigated area is located in the downstream countries of Sudan and Egypt, while rainfed agriculture is predominant in upstream catchments (NBI, 2012). However, many Nile riparian countries have plans for new irrigation developments. The basin has a huge hydropower potential. The potential hydropower in the Eastern Nile basin is more than 13,850 MW, of which 3,895 MW is currently operational through the main dams of Aswan High Dam, Sennar, Roseires, Jabel Aulia, Khashm Elgirba, Merowe and Tekeze. The hydro system of the Eastern Nile consists of ten major hydraulic infrastructures that are currently working as listed in **Table 1 - Appendix-I**. In Ethiopia, the series start with the Tana-Beles Scheme, which consists of an artificial link between the Beles River, a tributary of the Blue Nile, and Lake Tana, the source of the Blue Nile, to generate hydroelectricity (460MW) and to irrigate around 150,000 ha (planned). Then, the Tekeze dam is the largest hydraulic infrastructure in Ethiopia, with an installed capacity of 300 MW (Goor, et al., 2010). Only small scale irrigation exist in the in the Tekeze-Atbara river basin, but no large irrigation projects.

In Sudan, there are two major dams on the Blue Nile, Roseires (heightened by 10 meters in 2012, to double its storage capacity) and Sennar dams. The main objective of those dams is to regulate the seasonal flow of the Blue Nile waters for irrigation of more than one million ha of crops distributed over three irrigation schemes (Gezira, Rahad, Suki). Their electricity production is relatively small, attributed to the limited available head, 280 MW and 16 MW at Roseires and Sennar respectively. On the Atbara River, the Khashm Elgirba dam has a relatively small hydropower capacity (10.6 MW), and the new Upper Atbara Dams Complex completed in 2016. All abovementioned dams in Sudan face severe siltation problems. The siltation problem at Khashm Elgirba dam is managed by means of flushing. Reservoir sedimentation at Roseires and Sennar dams is managed by keeping minimum water levels during the flood season, and only start filling after the peak load of sediment has passed. Jebel Aulia dam, located on the While Nile near the confluence with the Blue Nile, provides water for irrigation schemes around the reservoir estimated at 275,000 ha. At the Main Nile, close to the 4th cataract, Merowe dam (12.5 x $10^9 \, m^3$) has an installed generation capacity of 1,250 MW and can potentially irrigate 380,000 ha.

In Egypt, there are five run-of-river dams and one major dam, the Aswan High Dam (AHD) being the major dam of the basin. The main objectives of AHD are to produce energy, to supply irrigation water, to regulate the flows to protect the downstream area against flooding and improve downstream navigation. The Old Aswan dam (OAD), located downstream of the AHD, is operated as a run-of-river hydropower plant. It is mainly used for hydropower production and to regulate the daily outflows from AHD (Goor, et al., 2010). The Esna run-of-river plant located downstream OAD is operated for hydro-power generation. The last three barrages, Assyut, Delta and Naga Hammadi divert Nile water to collectively irrigate 1.3 million ha.

Many new reservoirs and irrigation projects are proposed to be constructed in the Eastern Nile Basin, particularly in the Blue Nile sub-basin in Ethiopia as demonstrated in **Table-1, Appendix-I**. Not all proposed dams would probably be constructed due to a number of reasons, including financial obstacles, no strong market (for demand) in the region to use all potential hydro-electricity, and several reservoirs are proposed as alternative options.

2.4 COOPERATIVE PROGRAMS AND PROJECTS FOR WATER RESOURCES DEVELOPMENT WITHIN THE NILE BASIN

As is well-documented, a series of agreements over the utilization of Nile water were concluded during the colonial era and signed by Britain on behalf of most basin states (Allan, 1999). An important treaty after independence is the one signed between Egypt and Sudan in 1959, whereby these two countries allocated the mean annual flow of the Nile (84×10^9 m^3/yr) between them, namely 55.5×10^9 m^3/yr to Egypt and 18.5×10^9 m^3/yr to Sudan, while reserving 10×10^9 m^3/yr for evaporation losses from the Aswan High Dam. Reviews of treaties and agreements on the Nile basin are given in Salman (2013), Elshopky (2012), Fahmi (2007), Timmerman (2005), Dellapenna (2001) and Abate (1994). Different cooperative programs among the Nile countries (such as HYDROMET, UNDUGU, TECCONILE, FRIEND Nile, NRBCF, NBI) have taken place (Arsano & Tamrat, 2005; Demuth & Gandin, 2010; Hammond, 2013; Metawie & Sector, 2004; Salame & Van der Zaag, 2010; Wolf & Newton, 2013). A brief characterization of each programme (or project) is given in **Table 2.1**. Yet, so far there is no cooperative water management program encompassing all Nile riparian countries (Dellapenna, 2001; El-Fadel et al., 2003). This is probably attributed to the different interests of the eleven riparian countries and political instability in the region, and to the absence of regional institutions that govern water management issues in the basin. Water related issues are thus linked to the geopolitics of the basin (Abate, 1994; El-Fadel, et al., 2003; Sayed, 2008).

16

The Nile Basin Initiative (NBI) was established in 1999 to incorporate all basin countries through two major programs: Shared Vision Program (SVP) and Subsidiary Action Program (SAP). The NBI intended to provide a framework for basin-wide cooperation with the identification and implementation of new joint infrastructural projects (Goor, et al., 2010). In parallel, the Nile riparian countries embarked on a process to establish a permanent legal and institutional Cooperative Framework Agreement (CFA) (Mekonnen, 2010; NBI, 2010). However, only six countries (all located upstream) have so far signed the CFA while the two most downstream countries (Sudan and Egypt) did not sign as no consensus could be reached over one article in the agreement (Hammond, 2013).

The absence of a robust analysis of water resources development options in the Nile basin, and of the differential opportunities and risks these create for the riparian countries, may have contributed to the lack of consensus and mistrust among them (Subramanian et al., 2012). It is our belief that careful scenario analysis of cooperative management opportunities and risks can support basin or sub-basin wide cooperation on Nile water resources.

Table 2.1 Relevant cooperative programs and projects within the Nile basin with respect to water resources development

Activity Name	Date	Supported By	Geographic boundaries	Objectives
HYDROMET Project	1967-1992	United Nations Development Program (UNDP) and World Meteorological Organization	Equatorial Lakes and rivers	To collect, analyse and disseminate hydro-meteorological data to the Nile basin member states
UNDUGU (brotherhood) Forum	1983-1993	Initiative of Egypt	Egypt, Sudan, D. R. Congo, Uganda, and Central African Republic	Consultation on infrastructure, culture, environment, telecommunications, energy, trades and water resources
TECCONILE (Technical Cooperation Committee of the Nile Basin)	1992-1998	Canadian International Development Agency (CIDA)	Egypt, Sudan, Tanzania, Rwanda, Uganda and D.R. Congo	for Promotion of Development and Environmental Protection of the Nile Basin
NRBCF (Nile River Basin Cooperative Framework), D3 project	1995-1997	World Bank, UNDP, Canadian International Development Agency (CIDA)	All Nile basin states	To establish a regional co-operative framework for basin-wide integrated water resources planning and management.
FRIEND Nile Project (Flow Regime from International Experimental and Network Data)	1996-2015	UNESCO Flanders in Trust (FIT) Fund	All Nile basin states	To improve the Nile international river basin management through improving the cooperation among the riparian countries in the field of water resources management and

Activity Name	Date	Supported By	Geographic boundaries	Objectives
				regional scale analysis of hydrological regimes.
NBI (Nile Basin Initiative)	1999–2013	World Bank and other external partners	All Nile basin states	To achieve sustainable socio-economic development through equitable utilization of and benefit from the common Nile basin water resources.

Source: Arsano and Tamrat (2005), Hammond (2013), Metawie and Sector (2004), Salame and Van der Zaag (2010), Wolf and Newton (2013)

3

NILE RIVER BASIN MODELLING FOR WATER RESOURCES MANAGEMENT – A LITERATURE REVIEW [1]

The Nile basin water resources have been extensively studied during the last 125 years for planning and management purposes, in particular with regard to the use of blue water in the downstream part of the basin, though recently some studies have also focused on the upper parts. These studies show that there is no convergence of development plans emerging among the Nile riparian countries. This chapter reviews river basin water resource models as applied in the Nile River Basin, distinguishing between simulation, optimization and combined simulation and optimization models. The review aimed to identify knowledge gaps to guide future research on water resources planning and management in the Nile.

---

[1] This chapter is based on: Digna, R.F., Mohamed, Y.A., van der Zaag, P., Uhlenbrook, S.and Corzo, G.A., 2017. Nile River Basin modelling for water resources management – a literature review. *International Journal of River Basin Management*, 15(1): 39-52.

3.1 RESERVOIR-RIVER SYSTEM ANALYSIS MODELS

A river basin system consists of water source components and in-stream and off-stream demand components (McKinney et al., 1999). In addition to natural and physical processes, the river basin is characterized by development projects and management policies. River system planning and management is usually a multi-objective problem, with many objectives being in conflict (M. Karamouz, Szidarovszky, F., , 2003). The conflicts in river system planning and management arise when the water demands of different sectors are supplied from one river system and river flow is less than in-stream and off-stream water requirements. Infrastructures such as reservoirs are thus vital to organize and allocate the water for different water users.

A river can have a single reservoir with one or multiple objectives, or a cascade of reservoirs which are in series and/or in parallel. A system of parallel reservoirs occurs in a river that has many branches joint together in a junction point. Development of efficient operations can achieve substantial increases in benefits, however, reservoir system operation is complex. This complexity arises from many sources such as uncertainties in future inflows into the reservoir, demands, the trade-offs between wide ranges of conflicting objectives, and the relation among reservoirs in case of multiple reservoir systems. Therefore, there is no single type of reservoir operation problem, but rather a large number of decision problems and situations. Decision-making situations and decision support tools for reservoir operation can be categorized as pre-construction planning involving proposed new dams, post-construction planning involving re-evaluation of existing reservoirs operations, and real time operations (R. Wurbs, 1991).

Generally, system analysis models used to optimize reservoirs system operations are classified as: descriptive simulation models; prescriptive optimization models; and hybrid simulation and optimization models. Simulation models are useful for studying the operation of complex multiple reservoir systems that have relatively few alternatives to be evaluated. All values of operating decision variables should be defined before simulation can be performed. Optimization models are used to define a relatively small number of alternatives that can be tested, evaluated and improved by means of simulation (Loucks & Van Beek, 2005). Optimization of a multi reservoir system can be categorized based on method of computing inflow to stochastic and deterministic. Whilst the stochastic approach focuses on multi reservoir system evaluation through changes of inflow together with developing reservoir operating rules, the deterministic approach is concerned with how well optimization technique performs (T. Kim et al., 2006). Combining simulation and optimization models is often recommended for water resources planning and management. While optimization models can limit the range of feasible alternative scenarios, simulation models can analyse these alternatives to represent a realistic scenario system.

3.2 PREVIOUS STUDIES IN THE NILE BASIN

This section reviews the river basin modelling experience in the Nile basin, organized by modelling approach: simulation, optimization, and combined models. Each modelling approach start with a brief overview of the modelling concept highlighting advantages and limitations, after which for each sub-basin the respective studies are briefly described. The review aimed to identify an appropriate reservoir system modelling approach that achieves the research objective-1 (see Section 1.2).

3.2.1 River basin simulation models of the Nile

Water resources simulation models mimic the operation of the system given inflows, system characteristics and operating rules (Yeh, 1985). Some simulation models are not limited to the technical aspects, but also include social, economic and political issues. The performance of a simulated system can be assessed under certain conditions using hydrologic (i.e. in-stream flow, storage levels and generated hydroelectricity), economic (i.e. cost and benefit) and reliability indices (R. A. Wurbs, 1993). There are many advantages of water resources simulation models: they provide detailed and often realistic representations of the physical, environmental, economic, and social characteristics of the system (Simonovic, 1992). They are often flexible and relatively simple and provide insights into the dynamics and the structure of the system (Jacoby & Loucks, 1972). Simulation models are therefore widely used by water resources management agencies (Lund, 1999; R. Wurbs, 1991).

First, simulation models built for the entire Nile are reviewed, before we discuss models built for smaller parts of the basin. One of the first simulation model built for the whole Nile basin was the Nile Valley Plan (NVP), developed in 1958 by Morrice and Allan (1958). At that time the use of electronic computation and computers was very limited. Model runs were executed at the British ministry of defence in London, while other runs were executed in Paris. The NVP was designed to investigate the best controlling system for the Nile and its tributaries through the construction of dams (Morrice & Allan, 1958). Their study focused on the hydraulic aspects considering topography and hydrologic variability and used a trial and error approach to solve the water allocation problem. The windows based simulator (NILESIM) for the entire Nile river basin simulation was developed by the University of Maryland (Karyabwite, 2000). It was primarily built for educational purposes (Levy & Baecher, 1999). The simulation is limited to the hydrologic condition, other system conditions are not included in the model, such as future water needs, water quality and siltation problems.

Several river basin simulation models have been developed to study subsets of the Nile basin. In the Upper Blue Nile sub-basin an important hydro-economic study was conducted by the United States of Bureau of Reclamation (USBR) from 1958 to 1963.

23

The study investigated the potential of water resources development of the Ethiopian part of the Blue Nile basin for hydro-electric power generation and irrigation. The report was not limited to the hydrology, rather, it included groundwater, sedimentation, water quality, land use, geology, mineral resources, physiographic and local economy (Guariso & Whittington, 1987). The study concluded that the benefit/cost ratio for all major hydropower infrastructures on the Blue Nile main reach would be greater than 3.

A spreadsheet based hydro-economic model was developed for the Blue Nile downstream Roseires and Sennar reservoirs in Sudan to trade-off between sedimentation induced losses of storage volume and benefits from hydropower generation and irrigation. An operation policy for the two reservoirs system was also determined (Yasir Abbas Mohamed, 1990). The study used net present value (NPV) to evaluate different operating policies. Future irrigation expansion was however not considered in assessing the operating policies. A trade-off analysis between irrigation and hydropower generation for the same Blue Nile multi-reservoir system was conducted using a fuzzy set based stochastic simulation model (Abdallah & Stamm, 2013). The model was applied to find reservoir storages and releases using statistical flow, sediment and demand parameters. The study showed that the model performed well in compromising between the conflicting purposes of irrigation and hydropower generation.

Several studies developed simulation models for one (specific) natural or artificial reservoir (Abreha, 2010; Hurst et al., 1966; Wassie, 2008). Hurst (1966) suggested fixed monthly operation rules for the Aswan High Dam in an attempt to find the optimal operation that reduces the negative consequences. On the basis of simple routing calculations, the feasibility of the Toshka canal construction was investigated. Wassie (2008) used a generic simulation model, WAFLEX (Savenije, 1995), to assess the impact of different development scenarios on the fluctuation of Lake Tana water level and its outflow. Abreha (2010) used the RIBASIM model to quantify the trade-off between hydropower generation from Tekeze dam, downstream environmental flow, and new irrigation development in the western part of Ethiopia.

For the Eastern Nile sub-basin, simulation models including SWAT, RIBASIM and RIVERWARE have been developed under the Eastern Nile Planning Model Project (ENPM) (Belachew, et al., 2015). It appears that those models except for SWAT, and in a simplified way, did not address the siltation problem in reservoirs. Those models were presented in technical reports, and not widely published in scientific journals.

Kahsay et al. (2015) employed a Computable General Equilibrium (CGE) modelling framework using Global Trade Analysis Project model (GTAP) to estimate the direct and indirect economic impacts of the Grand Ethiopian Renaissance Dam (GERD) on the Eastern Nile economies. The study showed that the GERD would primarily benefits Ethiopia during the transient filling stage, while the benefits would extend to Egypt when the GERD gets operational.

24

The previous planning studies were based on the assumption that the historical climate conditions provide sufficient information for a reliable prediction of future system behaviour. While historical stream flows at the basin-scale have been relatively stationary, multi-decadal trends are evident at sub-basin scale (i.e. Blue Nile basin) (Meron Teferi Taye et al., 2015). The inter-annual variability of precipitation and thus streamflow of the Blue Nile sub-basin is influenced by the El Nino-Southern Oscillation (ENSO) climate phenomenon (Beyene et al., 2010; Meron Teferi Taye, et al., 2015). Zaroug et al. (2014) concluded that 83% of El Nino events starting in April-June lead to drought in the Blue Nile upper basin, while occurrence of extreme floods has a 67% chance when a La Nina event occurs immediately after an El Nino event. A number of studies have also indicated the sensitivity of Nile water resources to climate change (Conway, 1996, 2005; U. Kim et al., 2008; M. T. Taye et al., 2011). In general, different climate models gave different results on the impact of climate change on Nile water resources. Reviews of future climate and hydrology of the Nile basin and the Blue Nile sub-basin are provided by Di Baldassarre et al. (2011) and Taye et al. (2015), respectively. The impacts of climate change on water resources development in an economical context have been investigated by many scholars (Jeuland, 2010; Jeuland & Whittington, 2014; Matthew P. McCartney & Menker Girma, 2012). Jeuland (2010) applied a hydro-economic simulation framework to large infrastructure projects proposed in the Blue Nile (Karadobi, Beko Abo, Mandaya and Border) with a view to integrate climate change impacts into the economic valuation of these projects. The framework links three models; i.e. stochastic stream flow generation, hydrologic simulation and economic appraisal, to estimate the benefits of these projects in terms of their net present value (NPV). The stream flow generation model used historical stream flow and pre-processed runoff projections from the A2 climate change scenario to synthetically generate sequences of stream flow using a multisite autoregressive model. The study concluded that the benefits of the new Blue Nile projects under climate change are higher compared to the historical conditions. The study assumed that the operation of the planned dams is mainly for hydropower generation while maintaining minimum flows. The framework helps to assess the risks of infrastructure projects under different climate scenarios; however, the framework cannot be used to develop the operating rules of these infrastructures to reduce the associated risks. A similar study was conducted by McCartney and Menker Girma (2012), who assessed the performance of existing and planned irrigation and hydropower developments on the Blue Nile in Ethiopia under a midrange climate change scenario (A1B), while assuming no change in land use. Three models were applied in the study: a dynamic regional climate model COSMO-CLM, a rainfall-runoff SWAT model, and a water resources WEAP model. The outputs of the dynamic model (rainfall, temperature and evapotranspiration) were used as inputs into the hydrologic model to generate river flows and groundwater recharge, which were used along with the existing and future demand as inputs into the water evaluation and planning model. The boundaries of the study were the upstream Blue Nile up to the border between Sudan and Ethiopia. The study showed that the construction of large dams will contribute largely to the short-to-medium term economy because water availability for hydropower and irrigation will increase. Counter to Jeuland (2010), the study demonstrated that climate change would negatively influence the performance of large infrastructures and the economy. According to the study, the flow at the Sudanese-Ethiopian border would decrease as a result of climate change and upstream water resources developments.

The impact of climate change on the operation of the Blue Nile cascade of dams, Beko Abo, Mandaya and Border was also evaluated by Wondimagegnehu and Tadele (2015). They combined a global (ECHAM5) and regional climate model (RCM) of the A1B climate scenario, with a hydrological model, HEC-HMS, and a reservoir system simulation model, HEC_ResSim, to simulate current and future inflows and hydropower generation. The results indicate that the maximum and minimum temperature would increase, as would evapotranspiration, but precipitation would fluctuate. The results also showed that the annual inflow into Beko Abo and Mandaya would increase due to climate change, but the inflow into Border dam would decrease. Hydropower generation showed a similar trend of increase at Mandaya and Beko Abo and a slight decrease at Border.

In the same context and within the same domain of the Blue Nile river system in Ethiopia, Jeuland and Whittington (2014) applied a methodology for water resources development planning and operating strategy under the uncertainty of climate change through addressing the change of hydrologic parameters (i.e. temperature, precipitation and evaporation) ranging from -15% to +15%. The proposed analysis framework combined a hydro-economic simulation model that links system hydrology and climate change with a sensitivity analysis using Monte Carlo simulations to assess the variations of economic outcomes for different combination of new dams with the change of river flow. The new dams considered included Karadobi, Beko Abo, Mandaya and Border (in some alternatives as GERD). The study concluded that the alternative with a cascade of three dams (Beko Abo, Mandaya and Border) was economically most attractive. The study also showed that including GERD in any alternative made it economically less attractive because of the high capital cost and lower economic returns of GERD compared to Beko Abo as an initial investment. According to the study, GERD can be a feasible alternative when no more than two dams can be built, keeping water abstraction upstream low and in case of flows increase due to climate change.

Filling of the Upper Blue Nile cascade of dams, Karadobi, Beko Abo and Mandaya, were studied by Mulat and Moges (2014) using MIKE BASIM to develop filling policies that minimize the impacts on hydropower generation of GERD and the heightened Roseires dams and irrigation downstream. The model showed that filling duration of three, one and two years for Karadobi, Beko Abo and Mandaya, respectively would increase the maximum mean cumulative annual hydro-energy by 26% of the base scenario, in which

GERD was assumed online. The study also demonstrated that the irrigation supply in Sudan would not be affected as the supply reliability would remain greater than 80%. The study did not consider irrigation expansion downstream, however. The findings illustrate that impounding water upstream in the cascade of dams would reduce evaporation from GERD and Roseires reservoirs, implying that water levels at downstream reservoirs would reduce. MIKE BASIN was also used to assess the impacts of filing and operation phases of GERD on the performance of Aswan High Dam (Asegdew G. Mulat & Semu A. Moges, 2014). The study showed that a filling duration of six years would reduce the power generation by 12% of its current condition, while power generation would reduce by 7% after GERD would get operational. The results also showed a slight negative impact on the irrigation demands. The study assumed that no expansion on agriculture would take place during the filling stage.

Wheeler et al. (2016) tested various filling strategies for GERD and reoperation of downstream reservoirs using RiverWare and assuming coordinated operation of GERD. The study concluded that adaption of downstream reservoirs operation and basin-wide cooperative agreement can manage the risk to downstream users. A similar conclusion was reached by Zhang et al. (2016) through assessing the inter-annual and decadal-scale stream flow variability and different filling strategies of GERD.

King and Block (2014) evaluated the impacts of different filling policies for GERD under a range of climate scenarios. A tool combining a rainfall-runoff hydrological model with a reservoir operation and hydropower generation model was used to examine the performance of the dam and the resulted downstream flow. Future precipitation scenarios ranging from -20% to +20% were used to generate 100 stochastic time series of 50 years each. Five filling rate policies were assumed: hedging 5, 10, and 25% of stream flow entering the reservoir, retaining all amounts exceeding the historical average stream flow at dam site and retaining all amounts exceeding 90% of the historical average stream flow. The study showed that hedging 25% of incoming stream flow is found to be superior for filling time and generated hydropower, as it resulted in the highest cumulative power generation and the shortest period of filling (141 to 131 months corresponding to -20% and +20% change of precipitation, respectively). Conversely, 5% hedging policy would cause the least reduction of average downstream flow.

The literature showed that simulation modelling approaches have been used for assessing the system performance by means of scenarios. Different aspects of water resources in the Nile basin have been studied, giving most attention to climate change issues and water resources development planning. Climate change analysis was handled through using historical records and different climate change scenarios. Climate change studies showed different results of the future impacts of climate on the planned infrastructures and their economic feasibility. This is attributed to large uncertainty of predicted precipitation under climate change conditions. Small precipitation changes may result in large changes in river runoff. Alternative filling policies of new dams have also been investigated.

3.2.2 Optimization models of the Nile

Optimization models are mathematical models used to find the best way to meet different objectives of reservoir system management (Simonovic, 1992; R. A. Wurbs, 1993; Yeh, 1985). They are also used for planning purposes and real-time operation. In large basins that offer a variety of development opportunities, the number of alternative system plans can be extremely large; here optimization models could screen all alternatives to generate a limited number of feasible alternatives. Among the advantages of optimization models is their ability to incorporate values of social and economic variables in water resources allocation. Optimization models are normally not based on detailed system representation, a simplification which make them less suitable for the evaluation of system performance. Water resources system analysis using optimization has been given more attention by scholars in academia and system operational firms (C. M. Brown et al., 2015b; R. A. Wurbs, 1993; Yeh, 1985). The advances and improvement of optimization algorithms have enhanced the confidence in the findings required for policy makers and sustainable system management. However, reservoirs system optimization needs more attention as it is location-specific and depends on the scale of the analysis (C. M. Brown, et al., 2015b).

Optimization techniques used include classical optimization or mathematically based techniques such as Linear Programming (LP), Nonlinear Programming (NLP), Dynamic Programming (DP), and computational intelligence techniques or Heuristic Programming such as evolutionary computations, Artificial Neural Networks (ANN) and Fuzzy set theory (Labadie, 2004; Rani & Moreira, 2010). Optimization approaches can also be categorized based on the flow data used: stochastic and deterministic (Philbrick & Kitanidis, 1999; R. A. Wurbs, 1993). Stochastic approaches use statistical properties to generate series of flow data, whereas deterministic approaches use historical series of flow data. Stochastic optimization approaches have the ability to more realistically deal with uncertainties of future inflows than deterministic approaches. Côté and Leconte (2015) have, however, demonstrated that the use of deterministic scenarios is more robust than explicit stochastic optimization (probability distribution) in stream flow forecasting. Stochastic optimization cannot be applied to complex multi-reservoir systems that have a large number of variables, while deterministic optimization can be applied without simplifications (Philbrick & Kitanidis, 1999). Deterministic optimization can be applied for systems with no inequality constraints, linear system dynamics, quadratic performance function and the stream flows are dependent and normally distributed (Celeste & Billib, 2009; Philbrick & Kitanidis, 1999).

In the Nile basin, a number of optimization studies have been carried out. To evaluate the feasibility of Toshka canal project in Egypt, (Guariso, et al., 1981) developed a non-linear optimization model for the real-time management of the AHD. The analysis of the over year operation problem was compared with fixed operating rules as suggested by Hurst. A real-time management model proved superior in yielding a better hydrograph simulation. The study concluded that the construction of the Toshka canal was not feasible, even if the Jonglei canal project in South Sudan would be completed. Stedinger et al. (1984) demonstrated that both stationary and non-stationary stochastic dynamic programming models can identify better operation policies of AHD reservoir when improved hydrologic state variables are employed (such as using forecast equations for current period's inflow). In the same context, Kelman et al. (1990) developed sampling stochastic dynamic programming that differs from the conventional stochastic dynamic programming in that the complex spatial and temporal characteristics of stream flow processes can be captured from a large number of stream flow sequences.

Guariso and Whittington (1987) extended the work of Guariso et al. (1981) to examine the implications for Egypt and Sudan of the development of the upper Blue Nile water resources proposed in the USBR study of 1964. The study used linear programming (LP) with the objective functions of maximizing the hydropower in Ethiopia, and maximizing the water supply for agriculture in Sudan and Egypt. The study showed that water for agricultural use in Sudan and Egypt would increase and hydropower generation would decrease in Egypt as a result of reduced storage in AHD. The Rahad and the Dinder flows would also decline as the result of irrigation projects proposed by the USBR in Ethiopia. The model used many assumptions and approximations. The proposed four dams were represented by a single reservoir with their combined capacities; the hydraulic head on the turbines was not included; and the evaporation from all reservoirs was considered the same while actually it was not. The issue of sedimentation was not included in the study.

The results of Guariso et al. (1987) were in agreement with those obtained from the Nile economic optimization model (NEOM), the first economic model of the Nile basin developed by Whittington et al. (2005), which showed that the majority of irrigation benefits would be generated in Sudan and Egypt. NEOM was developed to optimize the entire Nile basin water resources in terms of hydropower generation and irrigation, considering the development of the four USBR proposed dams. The model was a deterministic non-linear (NL) constrained optimization model in which water resources network was presented as a series of nodes (reservoirs and irrigation schemes) and links. The economic benefit analysis was conducted at two levels: basin-wide level and country level. They concluded that the total economic benefits would almost be equal in Egypt, Sudan, Ethiopia, and the Equatorial States. However, the composition of the benefits differed between countries. For instance, the majority of the economic benefits of hydropower were generated in Ethiopia and to a lesser extent in Uganda. The model did not include the economic benefits and costs of flood control in the Nile, nor of reservoir sedimentation. Being a deterministic model, the model assumed a constant pattern of inflows in the future. The complexity of over year storage reservoirs cannot be addressed in this model (i.e. AHD) as it was an annual model which considers one year for the reservoir drawdown-refill cycle.

Goor et al. (2010) used a stochastic hydro-economic model to examine the mid to long term operation of infrastructures, in particular the four USBR proposed reservoirs in Ethiopia and the AHD in Egypt under normal condition of operation. Stochastic Dual Dynamic Programming (SDDP) was used for the optimization. Built-in multi-site-periodic autoregressive hydrological model parameters were estimated using flow data. The analysis showed that the flood peak level in the Blue Nile would decline because of flow regulation resulting from operation of the new dams. The hydroelectricity in Ethiopia would be boosted by 1666%, while the water allocated to irrigation in Sudan would increase by 5.5% and the inflows at AHD would reduce which was supported by the findings of previous economic studies (Whittington, et al., 2005). The model assumed that the operation of all infrastructures would be coordinated between the three riparian countries. The study did not explore the benefits foregone if the three countries would act unilaterally. Reservoir sedimentation was not included.

The same SDDP model was used to develop a hydro-economic model to assess the positive and negative impacts of GERD on Sudan and Egypt (Arjoon, et al., 2014). Similar to the results of Goor et al. (2010), the results showed that GERD would produce great benefits for Egypt, Sudan and Ethiopia from irrigation and hydropower in case the system would operate cooperatively. The results also demonstrated that GERD would play a significant role in removing hydrological uncertainty during the low flow period, although the model did not handle the inter-annual variability of the flow. The study also did not include the negative externalities that are expected from GERD such as the impacts on flood plain (recession) agriculture. Reservoir sedimentation was again not considered.

An economic evaluation for the cooperative and non-cooperative sediment management between upstream and downstream Eastern Nile basin for sediment control at AHD was studied using optimal control theory based on dynamic programming (DP) (Yoon Lee, et al., 2012). Optimal soil conservation upstream and determination of the best timing and technology for downstream reservoir sediment removal was considered in this deterministic model. Social benefits from cooperative management were larger than non-cooperative and baseline management scenarios.

In the Blue Nile basin two techniques based on the stochastic dynamic programming were used to derive the optimal operation policy of the Blue Nile double reservoir systems in Sudan (Ghany, 1994): sequential decomposition where each reservoir in the system was optimized and simulated separately, and conventional stochastic dynamic programming, in which the optimization for the two reservoirs was carried out simultaneously. The conventional stochastic dynamic programming gave better results compared to the sequential decomposition algorithm.

For the Blue Nile and Main Nile in Sudan Satti et al. (2014) used the General Algebraic Modelling System (GAMS) software to develop a deterministic hydro-economic optimization model to study the impact of upstream infrastructure developments and climate change. Climate change was considered by assuming a 20% increase and a 20% decrease of the river flow. The findings showed that the Blue Nile regulation resulting from upstream development would increase irrigation withdrawals only if upstream infrastructures induce a 50% decrease of the electricity price in Sudan, otherwise the optimal management would shift towards hydropower generation, including by downstream reservoirs such as Merowe. The study, however, did not include the planned agricultural expansion in Sudan nor the effect of upstream infrastructures on reducing sediment load.

An economic study confined to the upper Blue Nile ending at the Ethiopian-Sudanese border was conducted by Block and Rajagopalan (2007). They studied the economic benefits and costs of the potential hydropower and irrigation associated with the USBR proposed four dams during the filling stage and under climate change scenarios, using a hydrological model with dynamic climate capabilities (Investment Model for Planning Ethiopian Nile Development, IMPEND). Block and Strzepek (2010) used the same modelling framework for the economic valuation of the four proposed dams when ignoring the filling and sequencing of dams construction and for different climate scenarios, based on an analysis of historical records, ENSO and SRES. The study found that ignoring the filling stages of the four dams and assuming that all dams would be constructed simultaneously would overestimate the benefits by 6.4 Billion USD.

This review shows that most optimization methods used in the Nile river system analysis are mathematically based models. Similar to other river basins, simple optimization methods were used to analyse systems with more than two reservoirs. This is attributed to the large number of possible state variables. Other methods, such as data driven optimization, have proven to be capable of analysing complex systems (Asfaw & Saiedi, 2011; Rani & Moreira, 2010). However, these methods have so far not been used in studies on the Nile.

3.2.3 Combined simulation and optimization models of the Nile basin

Incorporating optimization techniques into a simulation model has proven to be a useful approach for the analysis of complex river basin systems (Loucks, 1979; Loucks & Van Beek, 2005; Rani & Moreira, 2010; R. A. Wurbs, 1993). The combined approach benefits from optimization in screening the full range of alternatives to generate the most feasible ones and from simulation in evaluating the response of the system generated by these

31

alternatives. The combined approach can be categorized according to the adopted mathematical model; the type of links between simulation and optimization modules such as: (i) use of simulation as sub-model of optimization model, (ii) imbedding optimization into simulation model, or (iii) the in parallel use of optimization and simulation; and to the operation rule that can be parameterized in simulation (Sechi & Sulis, 2009).

Few studies combined simulation and optimization models for the Nile river basin system analysis. Musa (1985) introduced a new approach to find optimal operation guidelines for the Blue Nile double reservoir multi demand system within Sudan. The approach combined river basin simulation model (MODSIM) with a heuristic optimization procedure and a linear programming model. Storages of reservoirs and shortages at demand sites were statistically analysed. The study concluded that the operation guidelines derived from the combined approach were improved over those obtained by using simulation alone. However, heuristic optimization cannot guarantee a global or local optimum solution to an optimization problem.

For the same reservoir system, a methodology for optimal operation during the dry season and trade-offs for the use of water for irrigation and hydropower generation was developed by Hamad (1993). An adaptive forecast simulation and optimization model (AFSOM) was used. The study combined four models, including: (a) an analytical model (REFORM) to forecast low flow, (b) a simulation model to determine the maximum area which can be planted during the dry season, (c) the IRIS model to calculate and forecast irrigation water requirements, and (d) a simulation and optimization model called Slice to optimize hydropower generation from the reservoir system. The combined model showed a considerable capability in enhancing decision making on hydropower generation and dry season cropping.

A simulation-based optimization model was used to develop a filling strategy for the proposed Mendaya reservoir on the Blue Nile in Ethiopia to minimize the impact on hydropower generation by the Roseires reservoir downstream in Sudan (Hassaballah, 2010). The multi objective optimization used a non-dominated sorting genetic algorithm (NSGA-11). The MIKE Basin simulation model was used to evaluate the optimization results for filling the Mendaya reservoir. The study concluded that NSGA-II was an efficient tool for solving optimization problems in a complex water resources system. A limitation of the study was that it focused on maximizing hydropower generation rather than the overall economic benefit of hydroelectricity production. Moreover, the future operation of Roseires reservoir was assumed to remain the same as the historical operation, which was mainly based on reducing the sediment during the flood. This condition is likely to change, not only after doubling its storage capacity (completed in 2012) but more so with the construction of upstream reservoirs.

The Nile River Basin Decision Support Tool (DST) was developed by Nile basin countries in collaboration with Georgia Water Resources Institute. It aimed to serve as a neutral tool for the use of the Nile countries to assess the benefits and trade-offs associated with different water development and management options. The Nile DST contains six main components (Andjelic, 2009): data base, river simulation and management, hydrologic modelling, agricultural planning, remote sensing, and a user-model interface. The model simulates the Nile response to different development, hydrologic, and management scenarios. Hydrologic watershed models and statistical procedures are used to generate stream flow forecasts where hydrologic models are unavailable. The optimization is based on the Extended Linear Quadratic Gaussian Control method (ELQGC). For river simulation, river and reservoir routing models are used to simulate the movement of water through the river (A.P. Georgakakos, 2007). DST was used to develop the Lake Victoria Decision Support Tool (LVDST) (A. a. Y. Georgakakos, H and Brumbelow, K and DeMarchi, C and Bourne, S and Mullusky, M, 2000), and the High Aswan Decision support system (A.P. Georgakakos, 2006) to support decision relates to water and energy resources in Uganda and to reservoir operation in Egypt, respectively. Georgakakos (2006) assessed the impact of the proposed four large hydropower projects on the Blue Nile on the flow and hydropower in the Nile basin using DST. The results showed a substantial increase of hydropower generation in Ethiopia, slight increase of hydropower generation in Sudan and a decrease of irrigation water supply deficit in Sudan. These results were conditioned on that consumptive water use would not increase in Ethiopia and Egypt, and that Sudanese and Ethiopian reservoirs were cooperatively managed. The study did not consider reservoir sedimentation problem and its potential impact on hydropower generation.

The Nile Basin Initiative (NBI) developed a new decision support system that improved the regional modelling system. The Nile Basin Decision Support System (NBDSS) encompasses three components: information system, analytical part containing simulation and optimization tools, and multi criteria analysis tool (Hamid, 2013). NBDSS uses stand-alone modelling tools for simulation and optimization, including the MIKE group of models and WEAP. The NBDSS is designed to run simulation and optimization models separately or in combination. The capabilities of NBDSS extended to model sedimentation, water quality and erosion. Hamid (2013) used NBDSS to assess the impacts of planned Ethiopian dams on the Sudanese reservoir system up to Khartoum. The study showed that the Blue Nile flow during the dry season would increase five times the current flow, along with significant social and environmental impacts resulted from losing floodplain agriculture and a reduced groundwater recharge rate.

Despite the advantages of the combined use of simulation and optimization method, there are few applications that outline the best management for the Nile basin. Most of the studies were carried out for the Blue Nile system and very few were conducted for the entire Nile.

3.3 STUDIES ON MANAGEMENT OF TRANS-BOUNDARY RIVER BASINS

Transboundary river water development can lead to conflict or cooperation between riparian (M. Karamouz et al., 2011; Madani, 2010; Rogers, 1969). Benefit sharing is suggested to resolve water conflicts (Dombrowsky, 2009a). However, Wu and Whittington (2006) argued that cooperation between riparian may not resolve water conflicts, if costs of cooperation are not justified by the scale of benefits. Therefore, determining and understanding the (wide) range of interrelated costs and benefits is key for improved management of trans-boundary rivers and thus better relations among riparian states (Sadoff & Grey, 2002).

The literature showed various conflict resolution techniques applied to shared river basins worldwide. Among these methods multi-criteria multi decision making approach based on conventional optimization methods is popular. Ryu et al. (2009) applied multi-criterion decision making approach to resolve water conflicts in Geum River Basin in Korea between upstream and downstream resulted from construction of two multi-purpose reservoirs.

Game theory is another conflict resolution method which based on multi-criteria decision-making approach. Madani (2010) argued that game theory results differ from those of optimization methods in such a way that in game theory each party tends to maximize individual benefits, contrary to optimization which assume cooperation towards maximizing the whole system benefits . He also indicated that despite the novel and usefulness of game theory, its integration into general system analysis is not yet well achieved.

Larijani (2009) developed cooperative game theory base method to resolve the conflict between hydropower generation companies and environment specialist in USA as a result of climate change impact on Federal regularity Commission. Game theory is also applied for conflict resolution in many basins such as Euphrates and Tigris (Kucukmehmetoglu, 2009), Sweden (Young et al., 1980), Nile Basin (Elimam et al., 2008).

Giordano et al. (2005) used a new approach for conflict resolution called cognitive mapping approach to develop a Community Decision Support System. Their work demonstrates that the system can assist in discussion and collaboration by helping participants to: formulate their problem, find out all available alternatives and the corresponding effects and constrains, and finally to identify their preferences. To assess performance, system is applied to a river basin in the south of Italy which has a problem of water allocation in dry periods indicating the potential of the approach with slight improvement of negotiation facilitation using fuzzy set theory, artificial intelligence and argument analysis.

34

The literature also showed other methods applied for conflict resolution such as watershed process simulation which applied to improve water regulation policies for hydropower and irrigation in Nepal (Pokharel, 2007), and graph model for conflict resolution applied to Northern America (Hipel et al., 2002) .

In the Nile basin, Wu and Whittington (2006) applied cooperative game theory to study the incentive structure of cooperative and non-cooperative plans for different states of the basin, through assuming partial and grand coalitions. Nile Economic Optimization Model (NEOM) is used to maximize the benefits of suggested coalitions.

3.4 RESERVOIR SEDIMENTATION ANALYSIS MODELS

Reservoir sedimentation has been intensively investigated and studied in the literature. However, a good understanding of many sedimentation processes is not yet achieved. Sedimentation processes prediction models are either over simplified or very complex, thus they cannot represent the reality (Garcâia, 2008; Sloff, 1991). Prediction methods for reservoir sedimentation processes are divided into empirical and mathematical methods.

Mathematical methods can include the interaction between hydraulic elements (i.e. energy equation, Manning's equation, and continuity principle), sediment movement, and boundary geometry (i.e. upstream boundary discharges, downstream rating curve, and storage-water surface elevation relation) (Morris & Fan, 1998; J. W. Nicklow & Mays, 2001). Compared to empirical methods, mathematical methods are more accurate and time dependent processes as well as the spatial behaviour of sediment and flow can more easily be analysed. However, full mathematical modelling for the processes is not desired as it makes the model very complex. Some processes are described empirically for the lack of knowledge. A lot of data requirements and inherent errors limit the reliability of the results (Sloff, 1991).

Trap efficiency (TE) method is an empirical method generally derived from records of reservoir sedimentation to give a quick approximation of the loss of storage capacity over time. TE is defined as the ratio of incoming and retained sediment load to the total incoming sediment load. The main parameters that the TE depends on are the watershed characteristics, water and sediment inflow and outflow, and reservoir storage (Kummu et al., 2010; Sloff, 1991).

Kummu et al (2010) developed a protocol to estimate basin-wide TE of the existing and planned reservoirs in the Mekong Basin based on Brune's method, in which TE is a function of residence time (equal to the effective reservoir volume divided by the mean annual discharge). Other factors that TE depends on, such as reservoir and dam type and sediment properties, are assessed through sensitivity analysis.

In most reservoir sedimentation models TE has either not been incorporated (i.e. assuming 100% TE) or been considered a constant value, time independent (Annandale, 2006; Minear & Kondolf, 2009). However, few studies have considered the dynamic aspects of TE. A dynamic TE was developed by Mohamed (1990) as a function of sediment particle fall velocity and rate of flow through the reservoir, assuming other factors such as reservoir operation method, type of outlet, age and shape of reservoir are negligible as they are constant for different operating policies. Lewis et al. (2013) modified Churchill equation to predict the daily sediment trap efficiency for the Burdekin Falls dam in Australia.

Minear and Kondolf (2009) incorporated both the change of TE with time and the impact of upstream reservoirs in trapping sediment in their model to iteratively calculate sediment yield using spreadsheet. They calculated TE based on Brown's method which depends on reservoir capacity and the drainage area. However, other variables that influence the TE were not considered, limiting the model when detecting regional trend and assessing the potential risk of sedimentation in reservoir is required.

Most of the techniques used to derive the optimal operation of reservoirs have not taken into account the loss of usable storage resulting from reservoir sedimentation. Few studies have paid attention to the relations between reservoir operations, sedimentation processes, and storage preservation (Schleiss, et al., 2016).

The relation between reservoir operations and sedimentation processes along the river and/or within the reservoir was investigated by many researchers (i.e. Carriaga and Mays (1995a) ; Nicklow and Mays (2000); Bringer and Nicklow (2001); Rashid et al. (2015); Bai et al., 2015).

Carriaga and Mays (1995b) developed a mathematical model to determine the optimum single reservoir releases that minimize sediment scour and deposition in downstream river reaches. Different optimization techniques (NLP solver, DP procedure, and DDP procedure) interfaced with a finite different modelling for sediment routing simulation. Reservoir sedimentation was not investigated.

A similar sedimentation problem but for a multi-reservoir network was investigated by Nicklow and Mays (2001). An optimal control model was developed to derive a multi reservoir release schedule that minimize sediment scour and deposition in both rivers and reservoirs using optimization-simulation interface. The sedimentation problem was formulated as discrete time optimal control problem and solved using successive approximation linear quadratic regulator optimization algorithm. The objective function was to minimise the cumulative change in river's and reservoir's bed elevations subject to operational constraints. All hydraulic and sediment transport relations are modelled in a finite different simulation model. Empirical trap efficiency based on Churchill method was used for reservoir sedimentation simulation. The methodology was mainly to solve the network sedimentation problem without considering the other purposes of the reservoirs. However, the model can be expanded to incorporate multi objectives covering the other aspects of reservoir operation.

Chang et al. (2003) investigated the relation between reservoir operation and storage preservation. A combined reservoir simulation and sediment flushing model was developed. GA was used in the reservoir simulation model to determine the optimal flushing operation rule curves. The sediment flushing model was used to estimate the amount of flushed sediment volume to update the elevation- storage relation. Khan and Tingsanchali (2009) developed reservoir optimization-simulation with sediment evacuation model. They used GA for rule curves optimization and simulation model for sediment evacuation modelling. An optimization-simulation with sediment evacuation model was also developed by Rashid et al. (2015) using GA optimization technique and Tsinghua equation to optimize the operation of multi-objectives multiple reservors system on Indus River in Pakistan.

A summary of different methods suggetsd in the literature for studing reservoir operation with sediment management are included in **Appendix-V**.

3.5 CONCLUDING REMARKS

This chapter reviewed a large number of modelling studies of the Nile River Basin for planning and management purposes, which are summarised in **Table 1 - Appendix-II**. A critical discussion was done by considering both modelling concepts and water related issues handled.

Simulation, optimization and combined modelling approaches have been used for both improving management of existing reservoirs system and for water resources development planning. Most reviewed studies focused on the latter. Despite that the Nile basin provides a wide spectrum of alternatives development options that cannot be fully investigated using simulating approaches, the use of simulation models in water resources planning analysis was dominant (half of the reviewed studies), compared to the other types of river basin models (36%, and 15% of the reviewed studies were optimization and combined models, respectively). Optimization methods were based on conventional methods such as mathematical, while data-driven and heuristic optimization approaches were dominant in combined simulation and optimization modelling approaches.

Various water resources development related issues have been investigated in the basin using river basin modelling, including

i.   the impact on downstream water availability of upstream developments;

ii.   the economic impact of water resources developments;

iii.   downstream impacts of alternative dam filling and dam operation policies; and

iv.   the impact of climate variability and change on water availability.

Although reservoir sedimentation is salient, sediment management has so far largely been ignored in the EN basin models, in particular in a multi-reservoir context.

This review was limited to river basin modelling studies that focused mainly on surface water. Since groundwater plays an important role in river basin management and its conjunctive use can, in cases, alleviate water stress in the basin, it is recommended to consider groundwater in future studies.

4

DEVELOPMENT OF THE EASTERN NILE SIMULATION MODEL USING RIBASIM [2]

In Chapter 3, it is mentioned that simulation methods provide detailed and realistic representations of the physical, environmental, economic, and social characteristics of the system. A river basin simulation model was developed for the Eastern Nile basin including twenty dams and twenty-one irrigation schemes.

In this chapter, the implications of water resources development in the Eastern Nile basin on water availability for hydropower generation and irrigation demands were assessed at country and regional levels, using scenario analysis methods. Sixty-four scenarios were used to test developments of several dams and irrigation demands, Grand Ethiopian Renaissance Dam (GERD) operation options, and unilateral (status quo) versus cooperative transboundary management of dams. The results show that water resources developments would have considerable but varying impacts for the countries.

---

[2] This chapter is based on: Digna, R.F., Mohamed, Y.A., van der Zaag, P., Uhlenbrook, S. and Corzo, G.A., 2018a. Impact of water resources development on water availability for hydropower production and irrigated agriculture of the Eastern Nile Basin. *ASCE Journal of Water Resource Planning and Management* 144(5): 05018007

4.1 INTRODUCTION

Water resources related issues in the Eastern Nile are complex (Belachew, et al., 2015). The river flow regime is characterized by large seasonal and inter-annual variability (Goor, et al., 2010). On the basis of source and use of water, the basin countries can be divided into two groups: the upstream countries of Ethiopia and South Sudan, which are net producers of Nile water and use relatively small amounts, and the downstream countries of Sudan and Egypt, which are net consumers of Nile water and use relatively large amounts of water. Most of the existing water resources developments in the Eastern Nile basin have taken place in the downstream part of the basin. The emerging upstream water resources developments would affect the existing downstream dams, leading to both positive and negative externalities.

Literature review in Chapter 3 showed that the Nile basin was modelled to address (specific) water resources related issues and associated implications, e.g., filling of planned dams, optimization of reservoir operation, impacts of climate change, etc. Different approaches were used (simulation, optimization, economic analysis, etc.), for varying topologies of the system, using different lengths of the boundary conditions. Although good insights of the system and expected impacts were given, still the picture is not fully understood for different topologies and probabilities of river inflows. Therefore, studying water resources development options in a regional context is still important to quantify the impacts both at regional and at country level. Quantifying benefits of managing the reservoirs system as one single unit, i.e., regardless of the political boundaries, is a prerequisite to quantifying potential benefits of cooperative management, which may stimulate cooperation among the riparian states.

In this chapter, the Eastern Nile water resources development options are quantitatively analysed, based on the recent plans for dam and irrigation development (2012), considering different management options. Four indicators are used: hydro-energy generation, irrigation supply reliability, evaporation losses induced by the reservoirs and the change of the basin's flow regime. A river basin simulation model for the Eastern Nile basin has been developed using RIBASIM. The analysis has been carried out through developing different scenarios for dam and irrigation developments, hydropower demands and system management options (**Table 4.1**). The scenarios have been run on a monthly time step for 103 years (1900 to 2002). The historical stream flows of the Nile basin have shown to be relatively stationary, though some trends are evident at localized tributaries (Meron Teferi Taye, et al., 2015). Taye and Willems (2012) demonstrated the occurrence of a multi-decadal pattern in the Blue Nile river. Therefore, use of a short data set of stream flow might be not sufficient. Unlike most previous deterministic and simulation-based studies, a long series of historical stream flow data has been used in the model to capture the temporal variability of flows. In addition, the use of RIBASIM simulation model facilitates a manual optimization of the scenarios through varying the sources of supply of the water users.

4.2 MATERIALS AND METHODS

4.2.1 Model and data

The Eastern Nile system (**Figure 4.1**) up to the Aswan High Dam (AHD) is modelled using a river basin simulation model, RIBASIM. The scenarios have been selected to represent the base case (S0), and then different dams' development in both Ethiopia and Sudan, as well as irrigation demands in both countries. Those reservoirs on the stem of main tributaries with high generation capacities and those irrigation projects with large demands for water are considered in this study (**Figure 4.1**). Six potential dam sites have been identified along the Main Nile in Sudan with a total potential energy generation capacity of 1,600 MW (Verhoeven, 2011). The potential of new irrigation in Sudan is estimated at 590,000 ha withdrawing water from the Blue Nile, 90,000 ha from the White Nile and 285,000 ha from the Atbara (ENTRO, 2007; W. N. M. Van der Krogt & Ogink, 2013). It should be noted that all current and plans for new irrigation development in Sudan on the Eastern Nile have water requirements that would exceed its agreed allocation with Egypt. Ethiopia's planned irrigation developments would further increase the pressure on water resources, in particular for Egypt. It is therefore unlikely that all planned irrigation developments would materialise.

RIBASIM simulates the performance of a system using hydrologic time series and allocation rules (Abreha, 2010; W.N.M. Van der Krogt, 2008; W.N.M. Van der Krogt & Boccalon, 2013; Verhaeghe et al., 1988). The model uses nodes and links to represent the river system components. The model links hydrologic inputs at various locations in the basin with water users. Water allocation can be simulated by setting source priority list for each water user. To allocate water among multiple competing demands, each water user has a specified water allocation priority. The monthly available water is allocated to the users by priority, first priority 1, next priority 2, etc. till the last specified priority. If users have the same water allocation priority then the upstream water users get the water before downstream users. As an example of the priority system of RIBASIM, water supply for the Gezira Scheme (abstracting upstream Sennar dam), is first supplied from Sennar dam, if not enough then from Roseires dam.

Data of the Eastern Nile basin has been collected from various sources, including: the Ministry of Water Resources and Electricity (MWRE) - Sudan, Nile Water Master Plan (MOI, 1979), Roseires Heightening Report (McLellan, 1987), periodical reports published by the Ministry of Agriculture - Sudan (Ministry-of-Agriculture, 2013) and data of the Eastern Nile Planning model (ENPM) from ENTRO (W. N. M. Van der Krogt & Ogink, 2013)

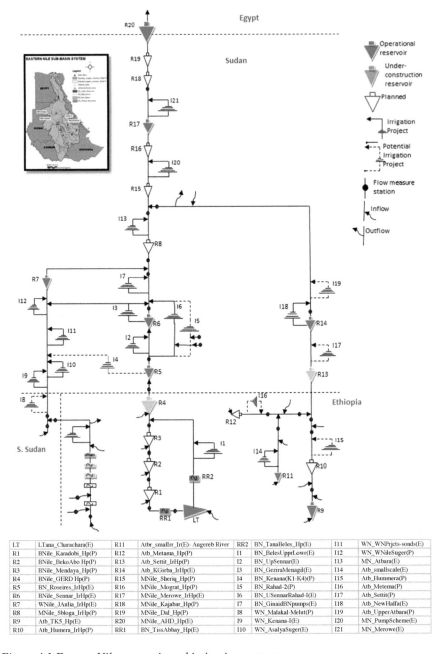

LT	LTana_Charachara(E)	R11	Atbr_smallrr_Ir(E)- Angereb River	RR2	BN_TanaBeles_Hp(E)	I11	WN_WNPrjcts-sonds(E)
R1	BNile_Karadobi_Hp(P)	R12	Atb_Metama_Hp(P)	I1	BN_BelesUpprLowr(E)	I12	WN_WNileSuger(P)
R2	BNile_BekoAbo Hp(P)	R13	Atb_Settit_IrHp(P)	I2	BN_UpSennar(E)	I13	MN_Atbara(E)
R3	BNile_Mendaya_Hp(P)	R14	Atb_KGirba_IrHp(E)	I3	BN_GeziraMenagil(E)	I14	Atb_smallscale(E)
R4	BNile_GERD Hp(P)	R15	MNile_Sheriq_Hp(P)	I4	BN_Kenana(K1-K4)(P)	I15	Atb_Hummera(P)
R5	BN_Roseires_IrHp(E)	R16	MNile_Mograt_Hp(P)	I5	BN_Rahad-2(P)	I16	Atb_Metema(P)
R6	BNile_Sennar_IrHp(E)	R17	MNile_Merowe_IrHp(E)	I6	BN_USennarRahad-I(E)	I17	Atb_Settit(P)
R7	WNile_JAulia_IrHp(E)	R18	MNile_Kajabar_Hp(P)	I7	BN_GinaidBNpumps(E)	I18	Atb_NewHalfa(E)
R8	MNile_Sbloga_IrHp(P)	R19	MNile_Dal_Hp(P)	I8	WN_Malakal-Melut(P)	I19	Atb_UpperAtbara(P)
R9	Atb_TK5_Hp(E)	R20	MNile_AHD_Hp(E)	I9	WN_Kenana-I(E)	I20	MN_PumpScheme(E)
R10	Atb_Humera_IrHp(P)	RR1	BN_TissAbbay_Hp(E)	I10	WN_AsalyaSuger(E)	I21	MN_Merowe(E)

Figure 4.1 Eastern Nile reservoir and irrigation systems

Table 4.1 Description of the scenarios

Developments	Country	S0 (Base, 2012 condition)	S1 (S10), (S11)	S2 (S20), (S21)	S3 (S30), (S31)
	Ethiopia	Tiss Abbay I, II	S0	S1	S2
		Tana Beles	GERD	Mendaya	
		TK5		BekoAbo	
		Atbr_smallrr_Ir(E)-Angereb River		Karadobi	
				Humera	
				Metama	
	Sudan	Roseires	S0	S1	S2
		Sennar	Roseires-Heightened	Settit	Dal
		Kashm Elgirba			Sheriq
		Jebel Aulia			Kajabar
		Merowe			Sbloga
Infrastructures (reservoirs)	Egypt	AHD	S0	S0	S0

Developments	Country	S0 (Base, 2012 condition)	S1 (S10), (S11)	S2 (S20), (S21)	S3 (S30), (S31)
	Total installed capacity (GW)	3.93	9.64	14.31	15.34
Annual irrigation water demand upstream of AHD (10^9 m^3)	Ethiopia	0	S10: 0 S11: 1.32	S20: 0 S21: 1.96	S30: 0 S31: 1.96
	Sudan	18.5	S10: 18.5 S11: 25.2	S20: 18.5 S21 :28.5	S30: 18.5 S31 :30.8
Annual water demand downstream of AHD* (10^9 m^3)	Egypt	55.5	55.5	55.5	55.5
	Total (10^9 m^3)	74	82.02	85.96	88.26

* The annual irrigation demand in Egypt is assumed to be equal to Egypt's water demand in the 1959 agreement (55.5 x 10^9 m^3/year).

To model the irrigation schemes of the basin, a fixed irrigation node was used. It requires data in the form of irrigated area (ha) and net average monthly demand (mm/day). In reality, the demands for most irrigation schemes (except those for perennial crops such as sugarcane) vary annually, as the cultivated area may be adjusted to fit the expected inflow. In this study, the demand (per ha) was assumed to remain constant over the years. The total potential area is used and assumed to be equally distributed between the different crops. Effective rainfall was considered negligible and ignored when determining irrigation demand. The potential areas of existing and planned irrigation projects in Sudan and Ethiopia have been taken from the Nile Water Master Plan (MOI, 1979) and from ENPM. Crop water requirement (ETcrop) (mm/day) of the potential and existing irrigation schemes have been calculated from FAO data including crop factors (Kc) and the Penman-Monteith reference evapo-transpiration (ETo) (mm/day). The total irrigation demand of Sudan in the base scenario thus amounts to 18.5×10^9 m³/yr. The annual irrigation demand in Egypt was assumed to be equal to Egypt's water demand in the 1959 agreement between Sudan and Egypt (55.5×10^9 m³/yr). The monthly demand pattern is taken from Oven-Thompson et al. (1982), the maximum monthly demand occurring during June and July. A similar assumption has been used by Goor et al. (2010) and Van der Krogt and Ogink (2013).

In RIBASIM, variable flow nodes are used to represent the natural water flowing through the river system. Water balance calculations are applied using a spreadsheet to generate the monthly time series of incremental natural flow of tributaries (represented by variable flow nodes) between gauge stations (record nodes). The hydrologic time series (103 years of monthly data set from January 1900 to December 2002) of the recorded (measured) station, rainfall and evaporation data at dam sites were supplied by ENTRO and as used in the ENPM. The model uses rainfall and evaporation data for the water balance calculations of the reservoirs. Effective rainfall data (1960-2000) are based on ERA40 gridded daily rainfall from the European Centre for Medium range Weather Forecast (ECMWF). Potential evaporation rates data of Egypt, Ethiopia and Sudan are based on the FAO database (W. N. M. Van der Krogt & Ogink, 2013). More details on data processing, generation and validation are available in Van der Krogt and Ogink (2013).

Model data of reservoirs in RIBASIM are the physical characteristics of the reservoir, main gate and hydropower plant characteristics (turbine capacity, efficiency, tail water level and losses), firm energy (demand and allocation priority) and operating rules. The operating rules are defined by identifying the flood control, target and firm storage levels and applying two hedging (reduction) methods for water releases from reservoir when water level drops below the specified firm storage level. Here, storage-based hedging was used. Storage-based hedging is supply based operation where reservoir releases are determined by the available storage and upstream inflow rather than the demand of downstream water users. Storage-based hedging requires defining distinct zones below firm storage and for each the percentage of the target release (full demand of all downstream users) that will be released for each zone (**Table 1- Appendix-III**); the lower zone from which water is released, the larger the reduction of the target release (W. N. M. Van der Krogt & Ogink, 2013). Operating rules of the planned dams are not known; we have chosen to simulate dam releases using the storage-based hedging method.

4.2.2 Simulation model

Two Eastern Nile models have been developed, one based on cooperative transboundary operation of all dams in the basin, and one where countries operate the dams unilaterally. This can be modelled in RIBASIM by settings in the source priority list. The list can either be empty or not. The default source priority list generated by RIBASIM model for each water user in a network includes all upstream supply sources that a user can receive water from. Water users with an empty source priority list cannot claim water from upstream sources to satisfy their demand and can only use the water available at their location, including uncontrolled flows (natural flows from variable flow nodes) and water released from upstream sources without considering downstream demand. A more detailed description of the water allocation procedure of RIBASIM is given in Van der Krogt and Boccalon (2013). For modelling cooperative transboundary management of the Eastern Nile system, the source priority list for each water user contains those upstream supply sources that can be used to satisfy the demand having the same logic of network links. In the unilateral scenario, the source priority list of the dams located near a border, i.e. Roseires, Khashm Elgirba (which is replaced by Settit dam once it gets online) and AHD were set as empty. The source priorities of the rest of the dams were not empty as there still is coordinated dam operation within each country; however, users cannot claim their demand from upstream sources beyond the border dam in their country.

Priorities of water users do not change with time but do with space depending mostly on the purpose of the supply infrastructure or dam. If the dam is constructed to be operated for hydropower generation only, such as the upstream Blue Nile dams in Ethiopia, generating firm demand will take priority over downstream demands. In case there is sufficient water to satisfy both firm energy and downstream water demands, such a reservoir releases water to fulfil all demands. In case water is insufficient, power generation takes priority over downstream demands and therefore the amount of water released for downstream demands will be reduced by the specified hedging rules.

If a dam is multipurpose for both hydropower and downstream irrigation, such as all existing dams, the priority will depend on the actual operation. For example, Roseires and Sennar on the Blue Nile of Sudan are operated for both hydropower and irrigation with the priority given to the irrigation demands of Sennar, Gezira and Managil schemes. For new dams with both hydropower and downstream irrigation dams such as Hummera and Settit on the Tekeze-Atbara River, hydropower and downstream irrigation were assumed to have the same priority.

The simulation cases within each model were compared to assess the implication of planned new dams and irrigation demands **(Research objective 1)**. The two models were also compared to assess the value of cooperative and unilateral operation for the dams in the entire basin and all countries **(Research objective 2)**.

4.2.3 Simulation cases

Apart from the baseline (S0), 12 scenarios were developed from the combination of (1) three dam development options (S1, S2 and S3); (2) two irrigation demand conditions; before any potential irrigation project realization (S10, S20, and S30), and after (S11, S21, and S31); and (3) two system management conditions: cooperative transboundary management with cases denoted as Sxx0, and unilateral management, with cases denoted as Sxx1 **(Table 4.1)**. Development of irrigation projects varies with scenarios because they are associated with the development of some dams that will be operated for hydropower generation and irrigation. The additional development of irrigation in S31 is attributed to development of irrigation schemes in the White Nile River; however, there are no planned dams on the White Nile. Operations of GERD are based on the uniform firm energy generation that can be satisfied 95% of the simulated time horizon. According to our simulations, the firm energy demand that GERD can satisfy is equivalent to 1,725 MW of continuous generation, while total energy generation reaches 15.1 TWh/yr, which is in line with Bates et al. (2012).

The baseline scenario (S0) considers the system as in the year 2011 before the heightening of Roseires reservoir. Data of the actual abstractions (e.g., for Gezira Scheme) are used to calculate the cropped areas A (ha) for model calibration and validation. In actual operation, the cropping areas of operational irrigation schemes in Sudan vary annually, based on the predicted inflow to Roseires dam; this is particularly true for the winter crops in central and northern Sudan. The average abstraction of irrigation projects per each month is therefore used to estimate the cropped area using given the monthly crop water requirement. The potential areas of irrigation projects are then used in the base and other scenarios.

The first scenario of dam development (S10) represents the system after GERD, and Roseires Heightening, with no additional irrigation development. The first scenario with irrigation developments (S11) includes additional irrigated agriculture in Ethiopia (total demand 1.32 x 10^6 m^3/yr), and in Sudan (total demand 25.2 instead of 18.5 x 10^6 m^3/yr). Therefore, the impact of GERD on the current system can be assessed by comparing scenarios S1x against S0. E.g., comparing S11 to S0 will indicate the impact of GERD on agriculture expansion of Sudan and also the impact of agriculture expansion on hydropower generation of the three countries.

The second scenario (S2) considers all dam developments upstream in Ethiopia at the Blue Nile and Tekeze-Atbara rivers **(Table 4.1)**, represented as S20 and S21 for no, and complete agriculture expansion, respectively. Therefore, comparing S2x to S0 will reveal the impact of upper basin full development on the hydropower and irrigation in the Eastern Nile system.

The third scenario (S3) represents full development of the basin dam and irrigation projects. S3 differ from S2 in that the Main Nile dams (S30) and irrigation projects (S31) in Sudan get online. Comparing S3 to S2 will indicate the impact of upstream and downstream water resources development on the basin's countries.

In the cooperative transboundary management scenarios, all water users are connected to one or more upstream sources depending on the network links. In case of two parallel reaches, water user located downstream the confluence will have two sources, the order of these sources depends on how much water each reach have. The most downstream demands are connected to the most upstream sources through the intermediate sources. For example, AHD demands can be fulfilled from its upstream source Dal dam, and Dal dam's demand from Kajabar dam, until the demand reaches Roseires and then GERD. When the system is managed unilaterally, the source priority list of AHD being empty, the demand of AHD cannot be fulfilled from Dal; rather, AHD receives only what Dal dam releases according to its own demand to produce energy (there is no irrigation demand between Dal and AHD). In other words, dams in each country are operated independently for the unilateral scenario, but could be dependently operated within the country.

4.2.4 Model assumptions

In this study, the current operation rules of all existing reservoirs are assumed to remain the same. All dam developments are assumed online and at operational stage; the transient stage (filling) and their short-term impacts have not been considered. In the initial condition of simulation, water levels of all reservoirs in the system are assumed full. The existing and proposed developments in Baro-Akobo-Sobat sub-basin have negligible effects on the system compared to the proposed large reservoirs in the other sub-basins and were therefore omitted. The potential irrigation projects of the upper basin withdrawing water from the Blue Nile and Tekeze-Atbara rivers are estimated at 0.2 x 10^6 ha (Goor, et al., 2010; W. N. M. Van der Krogt & Ogink, 2013). Domestic and industrial demands are negligible in the Eastern Nile basin compared to irrigation demand, therefore they were not considered. We further assume that the historical time series of 1900 to 2002 is representative of future discharges. This neglects any climate change effects, which is beyond the scope of this paper. Usable storage of the reservoirs was assumed to be constant in future, despite the fact that due to the siltation these storages are likely to reduce over time.

4.2.5 Model calibration and validation

For model calibration, the monthly irrigation demand was assumed to be identical to the measured abstractions of all irrigation projects during the year July 1970 - June 1971. The simulated abstractions of irrigation schemes and reservoir releases were compared to the measured ones.

Hedging rules based on storage, target levels of the operation rule and the power plant factor were used as adjustable parameters for calibration. The storage between firm level and dead storage level was divided into zones, water allocation at those zones were considered as a percentage of target releases and tested for different percentages between 100% and 20% resulting in significant improvement in the model output (**Table 1-Appendix-III**). The model was run for different target levels ranging between full reservoir level and firm level (or minimum operation) to adjust reservoir releases and supply of irrigation demand. As the power plant factor of existing dams of 90 % gave the best results, this factor was used. The results showed that the simulated and measured downstream releases and water levels of Roseires and Sennar dams are more or less the same. Also, the demand (measured) and supply (simulated) of irrigation projects are equal, indicating that the model performs well.

To reduce errors during model verification that could result from the change of available storage due to siltation, and thus resulting in differences between simulated and measured values, the physical characteristics of Level-Area-Volume relations of reservoirs derived from the available bathometric survey were adjusted according to the years of calibration and validation. Additional calibration data and results are provided in **Appendix-III**.

The model was validated using demand data for three years (July to June); 1977-1978, 1984-1985, and 1988-1989 representing normal, dry and wet years, respectively. For each hydrologic condition year, the model was run for the entire period (1900-2002) with the demand fixed at the actual abstraction of the year. The identification of the wet, dry and normal years was based on a comparison between the average monthly flow at Border (Eldiem) station 1965-2012 and the average monthly flow of the three years.

4.3 RESULTS AND ANALYSIS

Although results have been analysed for the 12 scenarios, the analysis focuses on the results of the scenarios that include GERD development under both cooperative transboundary and unilateral management, and with and without agriculture expansion. Other major results will be mentioned where relevant. However, the full set of results is available in **Appendix-III**. The section starts with presenting the validation results, then follow hydropower generation, irrigation development, and their impacts on evaporation losses from reservoirs and on the hydrographs.

4.3.1 Model validation results

Figure 4.2 displays the simulated and measured flow at the Blue Nile, and the Main Nile for a dry, normal and wet year. The results showed slight differences between simulated and assured flow during the wet season (July-October) downstream of dams in the Blue Nile River (**Figure 7-Appendix-III**). These differences are in part due to the filling and operation of Roseires, Sennar and Kashm Elgirba for sediment management. The time step used for filling (daily for 45 days) of Roseires and Sennar reservoirs differs from that used in the model (monthly). For reservoir sedimentation management, all gates are opened to release the coming inflow to pass the peak of sediment (and not to meet the downstream demands). The results also showed that simulated flow at Dongola station at the Main Nile is less than the measured flow, probably because of small flows from unmeasured tributaries of the Main Nile or to underestimated abstraction from the Main Nile.

The results of supplies and demands of Gezira, Managil and New Halfa irrigation projects during the three years showed that all the demands (the measured abstraction) are met, indicating the capability of the model to simulate the demand.

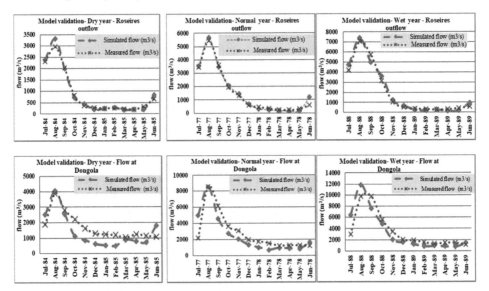

Figure 4.2 Measured and simulated flow at key locations in the Blue Nile, Atbara River and the main Nile at years of different hydrologic conditions: dry (July 1984-June 1985), normal (July 1977-June 1978) and wet (July 1988-June 1989)

The model accuracy was tested by calculating three model performance evaluation criteria: Root Mean Square Error (RMSE), Nash-Sutcliffe coefficient (E) and the correlation (r^2) for the simulated and measured stream flow at previously mentioned key stations. The results (**Table 4.2**) showed reasonable RMSE values (< half of measured flow standard deviation, according to Moriasi et al. (2007)) except at Khartoum, Tamanyat and Dongola station during the dry year. However; the correlation between simulated and measured flows at the two sites are very high (> 0.9) and Nash-Sutcliffe coefficients are reasonable (>0.5).

Table 4.2 Results of three measures for the model performance evaluation

Location	Dry year (July 1984-June 1985)			Normal Year (July 1977-June 1978)			Wet Year (July 1988-June 1989)		
	RMSE (m³/sec)	Nash-Sutcliff coefficient (E)	Correlation (r²)	RMSE (m³/sec)	Nash-Sutcliff coefficient (E)	Correlation (r²)	RMSE (m³/sec)	Nash-Sutcliff coefficient (E)	Correlation (r²)
Roseires	130 (< 1/2 CV)	0.93	0.97	205 (< 1/2 CV)	0.99	0.99	544 (< 1/2 CV)	0.97	0.98
Sennar	185 (< 1/2 CV)	0.92	0.97	294 (< 1/2 CV)	0.98	0.99	673 (< 1/2 CV)	0.95	0.96
Khartoum	475 (> 1/2 CV)	0.71	0.97	612 (< 1/2 CV)	0.93	0.96	1215 (< 1/2 CV)	0.86	0.89
Atb_K3	78 (< 1/2 CV)	0.99	0.92	147 (< 1/2 CV)	0.96	0.97	140 (< 1/2 CV)	0.98	0.99
Dongola	633 (> 1/2 CV)	0.81	0.91	1098 (< 1/2 CV)	0.92	0.98	1465 (< 1/2 CV)	0.91	0.91
Tamanyat	278 (> 1/2 CV)	0.89	0.96	455 (< 1/2 CV)	0.98	0.97	1020 (< 1/2 CV)	0.94	0.94

4.3.2 Hydropower generation

Figure 4.3 shows box-plots of the annual generated hydro-energy of the three countries for the base scenario (S0), and with GERD dam development (S1xx), including with/without irrigation developments (S10x, S11x), and cooperative transboundary/unilateral management scenarios (S1x0, S1x1). Hydro-energy generation in Ethiopia would boost by 1,500% after GERD gets operational (S100). Sudan hydrogeneration showed an increase of 17% (S100) compared to the present generation. Hydroenergy generation at AHD in Egypt would slightly decrease by 1% after GERD (S100). Despite the variation in the methodology and the downstream boundaries of the studies, the results have a similar order of magnitude as those reported by Arjoon et al. (2014) after GERD gets online; they found that energy generation would increase by 1,114% in Ethiopia, by 15% in Sudan and by 2% in Egypt. The fact that we find a slight decrease for Egypt can be explained by the possibility of operating AHD under relatively low water head level (Guariso & Whittington, 1987).

Figure 4.3 also displays the impact of irrigation developments on hydro-energy generation, where a general trend of reduction of energy-generation of the countries is shown compared to the without irrigation development scenarios. This is expected because of the consumptive nature of irrigation water. Energy generation in Sudan would reduce by 6.5% (S110), because most potential irrigation lies between Roseires and Sennar which both give priority to irrigation. The reduction in the case of AHD would reach 13% after upstream irrigation development (S110). The four scenarios for Ethiopia (S100, S110, S101, and S111) show no big difference. In other words, hydropower generation from the GERD is not affected by irrigation development – because the latter mainly occurs downstream. The overall basin hydropower generation is boosted by the GERD from 20,000 to over 35,000 GWh/year. This is not influenced by either cooperative transboundary or unilateral management, though slightly reduced by irrigation development.

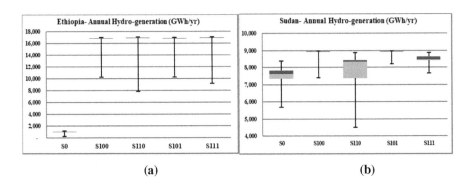

(a)   (b)

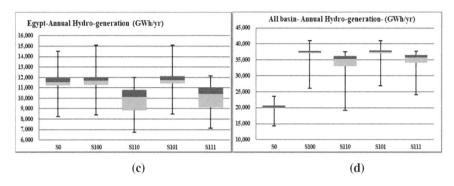

(c)   (d)

Figure 4.3: Box plot of the annual generated energy (GWh/yr) of the basin countries for each GERD dam development (S1xx) scenario, with (Sx0x) and without (Sx1x) irrigation development in case the system is managed in a cooperative manner (Sxx0) and unilaterally (Sxx1)

Table 4.3 Average annual generated energy at each country for irrigation development scenario, and cooperative transboundary and unilateral system management

Simulation case / scenario	Ethiopia		Sudan		Egypt	
	Coop (GWh/yr)	Non-Coop (GWh/yr)	Coop (GWh/yr)	Non-Coop (GWh/yr)	Coop (GWh/yr)	Non-Coop (GWh/yr)
S0	1,040	1,040	7,635	7,635	11,600	11,600
S10	16,865	16,865	8,951	8,951	11,526	11,768
S11	16,950	16,947	8,369	8,471	10,157	10,428
S20	35,260	36,034	9,273	9,081	11,777	11,698
S21	35,235	36,035	7,892	8,652	9,394	9,097
S30	35,260	36,034	15,220	15,074	11,875	12,064
S31	23,604	36,035	13,129	15,001	9,919	10,897

The results of considering additional hydropower dams (S2 and S3) are presented in **Table 4.3**. Although hydropower generation increases substantially by the new dams, all scenarios show no significant difference between cooperative transboundary and unilateral management except for S31. In the S31 scenario, Ethiopia hydropower generation reduces from 36,035 to 23,604 GWh/yr if the system operated in a cooperative fashion, while for Sudan (S310 vs. S311) hydropower generation reduces from 15,001 to 13,129 GWh/yr. Both reductions are attributed to the fact that in the cooperative case of system management, Ethiopian dams are operated considering the demand of the downstream countries, which has much increased because of the development of irrigation projects in Sudan; yet these demands would not be considered in the unilateral case. Similarly, the reduction of Sudan hydropower generation is because downstream demand of Egypt would be considered when operating the dams in Sudan, in addition to the increased demand resulting from the development of irrigation projects upstream the new hydropower dams of the Main Nile. Hydro-energy generation of Egypt would not much be affected by GERD, with or without cooperative transboundary management. This result is similar to that found by Arjoon et al. (2014), who show a negligible loss or gain in Egyptian hydropower generation resulting from unilateral management of the reservoir system (GERD). In the unilateral management scenario Egypt would nevertheless benefit from water released from the Merowe dam at the Main Nile for energy production, as this scenario (S111) does not yet consider irrigation expansion immediately downstream of Merowe.

4.3.3 Irrigation development

Table 4.4 summarizes the monthly supply reliability (average monthly supply to demand ratio) of existing and potential irrigation projects. The table shows a decrease in the supply-demand ratio of existing irrigation in Egypt by 1% after the GERD (S0 vs. S100 and S101), indicating no differences between cooperative transboundary and unilateral management of the system.

The reliability of irrigation supply to Sudan is practically not influenced by the GERD, but reduces by about 8% when upstream development and new irrigation expansion materialized. Cooperative transboundary management does not change results except for the last scenario S31, whereby reliability reduces from 90 to 80% from cooperative to unilateral management. For Ethiopia, reliability of irrigation supply significantly differs for cooperative transboundary and unilateral management (S11, S21, and S31).

The analysis of the probability of non-exceedance of irrigation supply of existing and potential projects in Sudan (**Figure 4.4**) reveals that the supply reliability of the existing irrigation in Sudan has a chance of 0.99 to be higher than 80%, in all scenarios and under both cooperative transboundary and unilateral management of the system, except in the case of full basin development and managed unilaterally; the chance would reduce to 0.75 (S301) (**Figure 9-Appendix-III**). A supply reliability of 80% represents an acceptable assurance of supply for irrigation schemes, given the possibility of practicing deficit irrigation (Steduto et al., 2012). Unilateral management of the system would not affect the chance of achieving a supply reliability of 80% for existing and potential irrigation with dams development except when all dams get online (S311) when it would reach 67% (**Figure 9-Appendix-III**). The supply reliability of irrigation projects in Ethiopia (not shown here) would be 1.00 for the scenario of GERD development under both cooperative transboundary (S110) and unilateral management (S111).

Table 4.4 Monthly irrigation supply reliability (average monthly supply to demand ratio (%)) of irrigation schemes in countries

Simulation	Cooperative transboundary system management			Unilateral management		
	Supply/ Demand ratio (%)			Supply/ Demand ratio (%)		
Case/scenario	Ethiopia	Sudan & S. Sudan	Egypt	Ethiopia	Sudan & S. Sudan	Egypt
S0	---	99	100	---	99	100
S10	---	100	99	---	98	99
S11	96	99	95	97	98	95
S20	---	97	100	---	98	99
S21	72	92	91	100	93	88
S30	---	96	100	---	93	99
S31	72	90	92	100	80	97

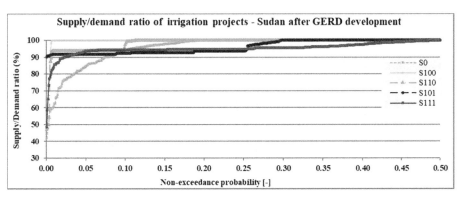

Figure 4.4 Non- exceedance probability of the average monthly supply to demand ratio (%) of Sudan existing(Sx0x) and potential (Sx1x) irrigation projects after GERD development (S1xx) under cooperative system management (Sxx0), unilateral management (Sxx1) and Base

4.3.4 Net evaporation losses from reservoirs

Figure 4.5 displays the average annual net evaporation from reservoirs of the countries at each dam developments scenario, with and without irrigation development, under cooperative transboundary and unilateral management of the system.

In case of cooperative management and without irrigation development, evaporation losses from Ethiopian reservoirs would increase from 0.20 x 10^9 m³/yr (S0) to 1.8 x 10^9 m³/yr after GERD is operational (S100). The average evaporation loss from Sudan reservoirs showed an increase to 6.2 x 10^9 m³/yr after GERD. Net evaporation from AHD would decrease from 13.3 x 10^9 m³/yr (S0) to 12.1 x 10^9 m³/yr after GERD (S100) gets operational, due to the reduced storage of AHD. Results in **Figure 4.5** indicate that, compared to the scenarios without irrigation development, the development of irrigation projects would induce small reductions of the net evaporation in Ethiopia and Sudan, and large reductions from Egypt's main reservoir, which is expected, because less water would be flowing into Egypt, resulting in AHD water levels to drop and with it the water surface area.

Taking a basin level perspective, the change of net evaporation from all dams would be insignificant after dam development in Ethiopia, while evaporation would increase with developments of the Main Nile dams. Unilateral system operation would have insignificant impact on net evaporation compared to that resulting from operating the system in a cooperative manner, until the development of the Main Nile dams, when net evaporation would increase as indicated in **Figure 4.5** due to the high evaporation losses in the Sudanese reservoirs on the Main Nile.

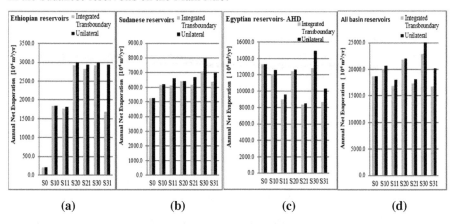

(a)   (b)   (c)   (d)

Figure 4.5: Average annual net evaporation from reservoirs under cooperative and unilateral management for the system, with and without irrigation development of: (a) Ethiopia, (b) Sudan, (c) Egypt, (d) entire basin

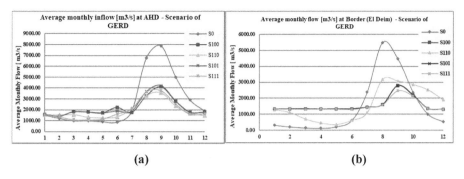

Figure 4.6 Average monthly flow [m3/s] at (a) Sudanese Egyptian border [Aswan High Dam (AHD)] and (b) Sudanese Ethiopian border [Border (Eldiem)] when GERD gets operational (S1xx), with existing (Sx0x) and potential (Sx1x) irrigation projects under cooperative (Sxx0) and unilateral (Sxx1) system management

4.3.5 Stream flow hydrographs

The average monthly inflows of the Main and the Blue Nile at the Egypt -Sudan (AHD) and Sudan-Ethiopia (Border or Eldiem) border are shown in **Figure 4.6**. The results show significant impacts of basin developments on the flow regime, represented by a reduction of the inflow during the wet season (July to September) and an increase during the dry season (October to April). In case of no irrigation projects are developed and the system is operated in a cooperative transboundary manner, the average monthly inflow at AHD would range between a minimum and a maximum of 1,420-4,135 m^3/s (average 2,186 m^3/s) after GERD (S100) compared to the base scenario (S0) (1,055-7,071 m^3/s with 2,733 m^3/s average). Development of irrigation projects would reduce flows to 1,239-3,570 m^3/s (average 1,915 m^3/s) after GERD (S110). The results are similar to the findings of Goor et al. (2010) and Arjoon et al. (2014) who also observed an augmentation of low flows and a reduction of high flows with GERD development. In case of unilateral system management, the variation would follow the same pattern, with a slight increase of the flow compared to those resulting from cooperative system management.

Inflows from Ethiopia at Border (Eldiem) would reduce in variability due to upstream dam developments. If the system is operated in a cooperative manner, the minimum and the maximum average monthly inflow would be 1,311-2,808 m^3/s after GERD gets operational (S100), compared to the base scenario (S0) (134-5,447 m^3/s). Unilateral system management would not significantly change these flows at Border (Eldiem).

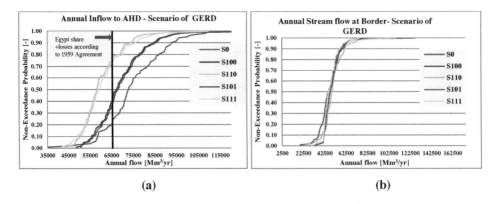

(a)   **(b)**

Figure 4.7 Cumulative distribution function (CDF) of the annual stream flow at (a) AHD and (b) Border when GERD gets operational (S1xx), with existing (Sx0x) and potential (Sx1x) irrigation projects under cooperative (Sxx0) and unilateral (Sxx1) system management

Figure 4.7 displays the probability of non-exceedance of the annual inflow at AHD and Border (Eldiem). According to the 1959 Nile water agreement between Sudan and Egypt, the inflow to AHD was supposed to be 65.5 x 10^9 m^3/yr, accounting for both Egypt's share (55.5 x 10^9 m^3/yr) and the additional evaporation losses due to the AHD that were then anticipated (10 x 10^9 m^3/yr). **Figure 4.7(a)** shows that the probability that Egyptian's claim is not met would increase from 23% in the base scenario (S0) to 42% if GERD (S100) would be in place and the system would be managed in a cooperative manner. The modelled probability of non-exceedance is relatively high in the base scenario compared to the generally accepted observations that AHD has so far mostly received annual inflow greater than the claimed share of Egypt. The high modelled value of probability of non-exceedance is because the model assumes that all irrigation schemes considered in the base scenario have been developed to their potential area, which is not yet the case.

The annual flow at the Sudanese-Ethiopian border (Border or Eldiem) shown in **Figure 4.7(b)** demonstrates that the probability of getting inflows greater than 48 x 10^9 m^3/yr is greater than 50% in the base case. The probability of getting the same inflow would remain the same in all dam development scenarios (S100, S200 and S300). When the system is operated unilaterally, the probability would not significantly change compared to the cooperative operation of the system.

4.4 CONCLUSION

A simulation model for the Eastern Nile basin was developed with which 12 scenarios (plus base scenario) were evaluated to assess the impact of dams and irrigation development in the basin based on four performance indicators: hydropower generation, irrigation supply reliability, evaporation losses from reservoirs, and change of the flow regime. The analysis focused also on the effect of system management, i.e., cooperative transboundary and unilateral management scenarios. The results of the simulation model indicate that dams and irrigation developments would generally have significant impact on the performance indicators.

The results indicated that compared to the current situation, the overall all hydropower generation of the basin would increase by 170% following hydropower dam developments in Ethiopian and cooperative management (S200). Results also showed that irrigation expansion with hydropower dams development in the basin, development of irrigation projects reduces the potential generated energy from the proposed hydro dams because of the consumptive nature of the irrigation. Similarly, full development of the proposed hydro projects would reduce the supply/demand ratios of irrigations schemes when all irrigation projects get developed, however, the supply/demand ratios are greater than 80% of the minimum crop water requirements.

The model provides quantitative information to understand the consequences of the available plans of dam development and agricultural expansion in the basin. The analysis does not include the influence of the high sediment load of some rivers (i.e. Blue Nile, Tekeze-Atbara) that significantly affects the usable storage of existing and future reservoirs. Further analysis of the silting up of reservoirs is required to better understand how dams affect and are affected by the sediment problem. In the Eastern Nile, sediment loads in rivers are a transboundary issue.

5

DEVELOPMENT OF THE EASTERN NILE OPTIMIZATION MODEL USING GENETIC ALGORITHM (GA)[3]

Optimal operation of multiple reservoir systems has been a subject of research for different water issues in different locations worldwide. Water resources system analysis using optimization methods allows to incorporate the economic values of water allocation between different water users and riparian in a transboundary river basin. This chapter presents the development and application of hydro-economic optimization model for the Eastern Nile basin. The chapter ends with the results obtained and conclusion.

---

[3] This chapter is based on: Reem F. Digna, M.E. Castro-Gama, Pieter van der Zaag, Yasir A. Mohamed, Gerald Corzo and Stefan Uhlenbrook, 2018b. Optimal operation of the Eastern Nile system using Genetic Algorithm, and Benefits distribution of water resources development. *Water* 10(7), 921

5.1 INTRODUCTION

Water resources system analysis, focusing on management strategies for sustainable and optimal use of water resources, can play an important role in conflict resolution by means of understanding conflicts and cooperation options. Water resources system analysis, in particular, multi-reservoir system optimization, has been given much attention by scholars in academia and system operational firms (C. M. Brown et al., 2015a; Chow & Cortes-Rivera, 1974; Kougias & Theodossiou, 2013; Larson, 1968; Mirchi et al., 2010; Murray & Yakowitz, 1979; Wardlaw & Sharif, 1999; R. A. Wurbs, 1993; Yeh, 1985). The advances and improvements of optimization algorithms have enhanced the confidence of policy makers in the search for sustainable system management. However, reservoir system optimization needs more attention as it is a location-specific and depends on the scale of the analysis (C. M. Brown, et al., 2015a)

The literature suggests various conflict resolution techniques applied to shared water courses. These techniques use multi-criteria decision-making approaches based on methods, such as conventional optimization from operation research, to more advanced ones, such as game theory. Madani (Madani, 2010) argued that game theory results differ from those of optimization methods in such a way that, in game theory, each party tends to maximize his benefits. This is in contrary to optimization, which assumes cooperation towards maximizing the whole system benefits. Nash equilibrium solutions can be applied in game theory to maximize the benefits of non-cooperation conditions between players. In the Nile Basin, game theory is applied to study various levels of cooperation and non-cooperation among the states of the basin (Dinar & Nigatu, 2013; Elimam, et al., 2008; Wu & Whittington, 2006).

In the context of system analysis, the Eastern Nile River system, with its many reservoirs, can be defined as having multiple objectives, predominantly for hydropower and irrigation, constrained by conflictive objectives, and high upstream–downstream interdependencies (R. F. Digna et al., 2017). Many scholars have applied different system optimization techniques to study the Eastern Nile River system, addressing the allocation of water from existing and planned dams among different users and riparian countries under different management options. These methods include mathematically based (conventional optimization) techniques, such as Linear Programming (LP) (A.P. Georgakakos, 2007; Guariso & Whittington, 1987), Nonlinear Programming (NLP) (P. Block & Strzepek, 2010; P. J. S. Block, Kenneth Rajagopalan, Balaji, 2007; Guariso, et al., 1981; Guariso & Whittington, 1987; Jeuland, et al., 2017; Satti, et al., 2014; Whittington, et al., 2005), Dynamic Programming (DP) (Arjoon, et al., 2014; Goor, et al., 2010; Habteyes, et al., 2015; Y. Lee, et al., 2012), and computational intelligence techniques (Hassaballah et al., 2011). Chapter 3 provided a comprehensive review on diverse Nile River Basin models and simulation techniques. The findings of these studies showed some discrepancies and common agreement on the impact of development of water resources infrastructures upstream in the Eastern Nile Basin on downstream hydropower generation and irrigation water supply. The results showed a common agreement that water availability for irrigation might increase and hydropower generation may not be affected or reduce slightly, while there is discrepancy on quantifying these impacts.

There are, however, some limitations of the application of conventional optimization techniques, in particular when they are used in a complex multi-reservoir system having hydropower generation as one of its main objectives. Linear optimization techniques are efficient for large-scale systems with high-dimensional variables, but require all relations among variables in constraints and objectives to be linear (F. Li et al., 2013; Loucks & Van Beek, 2005; Rani & Moreira, 2010). Though, it is not applicable for system analysis with inclusion of hydropower generation, without linearization and/or simplifications. Nonlinear Programming is effective for handling nonlinearity; however, it requires that all relations must be differentiable, which might not always be applicable for complex problems that have non-concave, non-convex, discontinuous and non-differentiable functions. Dynamic programming can handle nonlinearity in objective functions and constraints and continuity of the functions. However, dimensionality or handling multiple state variables is one of the dynamic programming limitations. The number of discrete combinations of state variables increases exponentially as the number of state variables increases. Evolutionary computation approaches, such as Genetic Algorithm (GA), overcome the limitations of conventional optimization techniques in reservoir system analysis, and deal with nonlinear, discontinuous, non-convex and multi-functions (John Nicklow et al., 2010). GA has been successfully applied worldwide for reservoir optimization (Rashid et al., 2015). GA has been found to be superior among other conventional methods in that it can get global or near global optimal solutions because of its search concept of population of solutions (Momtahen & Dariane, 2007). GA uses the operators for initialization, fitness, crossover and mutation to generate a multiple Pareto-optimal solution in one run for a multi-objective optimization problem. GA can save computation time when used for large-scale problems due to its parallel processing nature, in addition to the possibility of using the same computer code for different problems. However, GA is not appropriate for highly constrained problems because of the big portion of infeasible solutions, which may result in the population (Hakimi-Asiabar et al., 2010). Despite its robustness, evolutionary computation algorithms have not yet been applied in the complex Eastern Nile system.

The aim of this study is to analyse optimal scenarios for water resources management in the Eastern Nile with regard to hydropower generation and irrigation development **(Research objective 3 in section 1.2)**. A hydro-economic optimization model based on GA is developed to determine the maximum benefits for two scenarios **(Research objective 4 in section 1.2)**: (i) non-cooperative management of hydraulic infrastructures by the riparian countries, and (ii) cooperative water resources management among the riparian countries.

Application of GA in water resources problems is not new; however, specifically in a complex system, such as the Eastern Nile Basin, to the best of our knowledge, most approaches used before are single-objective oriented or based on diverse operation research methods. Such a deterministic optimization approach allows for the simultaneous inclusion of all hydro-dams and irrigation schemes, existing and planned without simplification, such as handling over-year storage. A deterministic optimization approach is recommended for complex systems, where large numbers of variables can be analysed without simplifications (Philbrick & Kitanidis, 1999).

5.2 MATERIALS AND METHODS

5.2.1 The Eastern Nile Optimization Model (ENOM)

To assess the distribution of benefits between the riparian countries from the optimal operation of the system under both cooperative and non-cooperative management, a deterministic hydro-economic optimization model for the Eastern Nile basin (ENOM) is developed. Hydro-economic models economically interpret the impact of water resources development and hydrological changes on the related water system and riparian states (Harou et al., 2009; Jeuland, et al., 2017) .The model has two components: (i) an optimization model and (ii) a river basin simulation model. **Figure 5.1** illustrates the conceptual framework of the ENOM. Both optimization and river basin simulation models are coded in MATLAB. The optimization model uses a Genetic Algorithm (GA) to optimize the water releases from reservoirs for hydropower generation and irrigation.

The ENOM is formulated to maximize the aggregated net benefits associated with water allocation for hydropower generation (f1) and irrigated agriculture (f2) by identifying optimal turbine release and irrigation withdrawal RIRt at each time step (t) over time horizon (T). The optimization problem is written as follows.

5.2.1.1 Decision variables

Decision variables (Rt) represent each reservoir releases through the turbines (Rw) and abstracts for irrigation (IR) at each time step (t). Rt is a vector of the following form:

$[Rt] = [Rw_{1,1}, Rw_{2,1}, ..., Rw_{T,1}; Rw_{1,j}, Rw_{2,j}, ..., Rw_{T,j}; ...; Rw_{1,J}, Rw_{2,J}, ..., Rw_{T,J}; IR_{1,1} IR_{2,1}, ..., IR_{T,1}; IR_{1,i}, IR_{2,i}, ..., IR_{T,i}; ...; IR_{1,I}, IR_{2,I}, ..., IR_{T,I}]$

The total number of decision variables (nvar) is equal to:

$$nvar = T * (J + I)$$

5.2.1.2 Objective function

The objective (F) is to find the combined reservoir releases and abstraction from reservoirs (RIRt) that leads to maximize the returns from hydropower generation (f1 and irrigation projects (f2) of the whole system during the time horizon (T). The objective function can be written as:

$$F(S_t, I_t, R_t) = \max_{RIR_t}\{f1, f2\} \tag{5.1}$$

$$f1 = P_e \sum_{t,j}^{T,J} HP_{t,j} \tag{5.2}$$

$$HP_{t,j} = C * \tau_{t,j} * \eta_{t,j} * H_{t,j}^{net} * R_{t,j} \tag{5.3}$$

$$f2 = P_w \sum_{t,i}^{T,I} IR_{t,i} \tag{5.4}$$

where:

Symbol	Unit	Description
$HP_{t,j}$	MWh/month	Total generated energy from Reservoir (j) at time (t)
P_e	US\$/MWh	The economic benefit of generated energy
C	N/m^3	Constant represents specific gravity and unit conversion
$\tau_{t,j}$	hours/month	Number of hours in period (t)
$\eta_{t,j}$	-	Turbine efficiency
$H_{t,j}^{net}$	m	Turbine Net Head of reservoir (j) at time (t)
$Rw_{t,j}$	m^3/month	Turbine discharge of reservoir (j) at time (t)
$R_{t,j}$	m^3/month	Release state variables from reservoir (j) at time (t)
P_w	US\$/m^3	The economic benefit of withdrawal water for irrigation
$IR_{t,i}$	m^3/month	Withdrawn water for irrigation (i) at time (t)
S_t	m^3/month	Storage state variable at time (t)
I_t	m^3/month	Inflow state variables at time (t)
T	month	Planning time horizon
J	-	Total number of dams in the system
I	-	Total number of irrigation schemes in the system

5.2.1.3 Constraints

The objective function is subject to the following constraints:

Energy generation constraints:

$$HP_{t,j} \leq HP_{t,j}^{max} \tag{5.5}$$

$$q_{t,j}^{min} \leq Rw_{t,j} \leq q_{t,j}^{max} \tag{5.6}$$

Reservoir storage limits:

$$S_j^{min} \leq S_{(t,j)} \leq S_j^{max} \tag{5.7}$$

Irrigation withdrawal limits:

$$IR_{t,i}^{min} \leq IR_{t,i} \leq IR_{t,i}^{max} \tag{5.8}$$

$$IR_{t,i}^{min} = \propto * (A_i * CW_{t,i}) , \qquad IR_{t,i}^{max} = (A_i * CW_{t,i}) \tag{5.9}$$

$$0 \leq \propto \leq 1 \tag{5.10}$$

Continuity (mass conservation) constraints:

$$S_{t+1,j} = S_{t,j} + I_{t,j} + C_{j,k}^R (Rw_{t,j} + Sp_{t,j}) + C_{j,z}^{IR} (IR_{t,i}) - e_{t,j} \tag{5.11}$$

$$e_{t,j} = A_{oj} * Ev_{t,j} + A_{tj} * Ev_{t,j} * (S_{t+1,j} + S_{t,j})/2 \tag{5.12}$$

$$Sp_{t,j} = S_{t+1,j} - S_j^{max} \quad if \ S_{t+1,j} > S_j^{max}$$

$$\text{Otherwise, } Sp_{t,j} = 0 \tag{5.13}$$

End-storage constraint:

$$\forall_j , S_{T,j} \geq D_j \tag{5.14}$$

Non-negativity constraints:

$$Rw_{t,j} , S_{t,j} , IR_{t,i} , HP_{t,j} \geq 0 \tag{5.15}$$

Additional constraint for Sudan's irrigation withdrawal from the Nile Agreement (1959), which identifies Sudan's share of the total Nile runoff:

$$\sum_{y=1}^{Y} \sum_{i_{su}=1}^{I_{su}} IR_{isu,t} \leq 18.50 \times 10^9 [m^3/year] \tag{5.16}$$

where:

Symbol	Unit	Description
$HP_{t,j}^{max}$	MWh/month	Maximum hydropower energy could be generated from reservoir (j) at time (t)
$q_{t,j}^{min}$	m³/month	Minimum turbine discharge of reservoir (j) at time (t)
$q_{t,j}^{max}$	m³/month	Maximum turbine discharge of reservoir (j) at time (t)
$S_{t,j}$	m³/month	Storage state variable of reservoir (j) at time (t)
S_j^{min}	m³	Minimum storage volume of reservoir (j)
S_j^{max}	m³	Maximum storage volume of reservoir (j)
D_j	m³	Target end storage of reservoir (j) at time (T)
$IR_{t,i}^{min}$	m³	Minimum water withdrawn for irrigation (i) at time (t)
$IR_{t,i}^{max}$	m³	Maximum water withdrawn for irrigation (i) at time (t)
A_i	m²	Irrigated area of scheme (i)
$CW_{t,i}$	m/month	Crop water requirement of irrigation scheme (i) at time (t)
$\propto$	-	Coefficient representing supply/demand ratio
$S_{t+1,j}$	m³/month	Storage state variable of reservoir (j) at time ($t + 1$)
$I_{t,j}$	m³/month	Inflow state variables at reservoir site (j) at time (t)
$Sp_{t,j}$	m³/month	Spillage of reservoir (j) at time (t)
$e_{t,j}$	m³/month	Evaporation loss of reservoir (j) at time (t)
A_{oj}	m²	Surface area of reservoir (j) at the dead storage level
A_{tj}	m²/m³	The area per unit storage of reservoir (j)
$C_{j,k}^R$	-	Reservoir system connectivity matrix = −1 when abstraction, +1, receives water from upstream reservoir [reservoir (j) receives water from reservoir (K)]
$C_{j,z}^{IR}$	-	Irrigation system connectivity matrix = −1 when abstraction, +1, receives return water from upstream irrigation [reservoir (j) receives water from irrigation (i)]

Two functions are performed in the optimization model of the ENOM; computing the fitness values (objective function) for each set of decision variables, and generating reservoir releases (decision variables) for hydropower and irrigation. First, parameters of GA operators are selected, such as population size and creation functions (constrained and unconstrained), numbers of generations, selection, mutation and cross over methods, and termination criteria. The GA generates sets of populations. At each generation, sets of decision variables (releases) forming a population are randomly generated between upper and lower bounds based on **Equations (5.6)** and **(5.8–5.10)**. The fitness values are then computed for each set of decision variables (**Equations (5.1), (5.2) and (5.4)**) and ranked; the sets with high scores are kept for the next generation. Releases are used in the river basin simulation model to compute reservoir storages, water levels and generated energy, based on **Equations (5.3), (5.11–5.13)** and (**Figure 5.1**). The termination criteria are checked following evaluation of fitness values; the model stops if the criteria are satisfied, otherwise, the next generation continues with new generated sets of decision variables and those carried from the previous generation with high scores. The process continues evolving towards optimal solution till the termination criteria are satisfied.

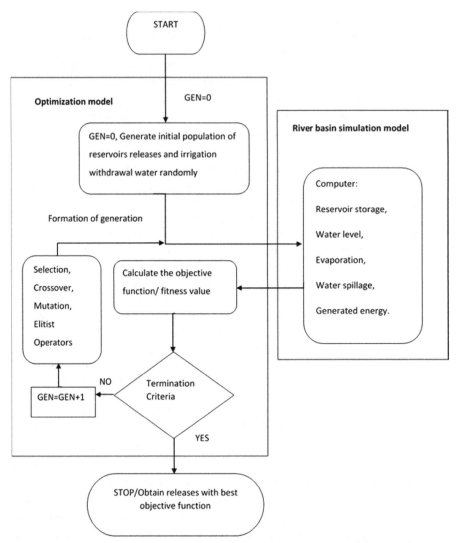

Figure 5.1: Conceptual framework of Eastern Nile Optimization Model (ENOM)

To overcome the GA limitations on handling the highly constrained system, the nonlinear constraints are satisfied in different ways to transform the constrained optimization into the unconstrained one. The computation of reservoir storage in the river basin simulation module is based on the continuity equation; therefore, continuity constraint is satisfied. Storage and end-storage constraints are included into the objective function in form of penalty functions. The deviation from the minimum and maximum storage and end-storage are penalized by square differences from constraints limits as:

$$\sum_{t,j}^{T,J} c_1 \left(\min \left(0, \left(S_j^{\min} - S_{t,j} \right) \right) \right)^2 \qquad (5.17)$$

$$\sum_{t,j}^{T,J} c_2 \left(\min \left(0, \left(S_{t,j} - S_j^{\max} \right) \right) \right)^2 \qquad (5.18)$$

$$\sum_{j}^{J} c_3 \left(\min \left(0, \left(D_j - S_{T,j} \right) \right) \right)^2 \qquad (5.19)$$

where c1, c2, c3 are constants, representing the weight of the penalty terms in the objective function.

The ENOM runs on a monthly time step. ENOM allows assessing different system management and water availability conditions. The model can optimize the whole system as one unit or per country to represent the cooperative and non-cooperative system management condition. It also has an extended module to simulate sedimentation in reservoirs using the trap efficiency method. All reservoirs on the stem of the main rivers of the basin were modelled; those developed on the small tributaries were not considered. The simulation network (**Figure 5.1**) includes 20 existing and planned dams: 6 dams on the Blue Nile reach (4 planned dams on the Ethiopian Blue Nile reach and 2 existing dams on the Sudanese part), 1 dam on the White Nile in Sudan, 6 dams on the Tekeze–Atbara River (4 dams in the Ethiopian part and 2 dams in the Sudanese part of the river), 7 dams on the Main Nile (6 in the Sudanese part and 1 in the Egyptian part), and 21 irrigated agriculture schemes representing existing and planned developments in Sudan and Ethiopia. The total water storage capacity of the system is approximately 341×10^9 m³ to irrigate an area of approximately 3 million ha (**Figure 5.1**). The downstream boundary of the simulation network is AHD. Irrigation demands of the downstream AHD are assumed as 55.5×10^9 m³/yr (Oven-Thompson, et al., 1982), equivalent to Egypt's water demand according to the 1959 agreement between Egypt and Sudan, due to data limitations.

For the purpose of this study, the model was run at a monthly time step, and included only 9 reservoirs and 14 irrigation schemes, representing the existing system as well as the Grand Ethiopian Renaissance dam (GERD) in Ethiopia, which is under construction (**Table 5.1**). The analysis covered system optimization to satisfy the demands of the main users in the basin, irrigation and hydropower; it did not cover other impacts of system optimization on reservoir sedimentation, environmental criteria or flood control. The ENOM was not intended for real-time or operational purposes. The operation we attempted to optimize was mid-to-long term operation. For planning purposes, the monthly time step was quite fair, especially in case of over-year storage reservoirs.

The data used in the simulation model were obtained from the previous chapter (Digna et al., 2018a). The key input data were the physical characteristics of dams, stream flows, evaporation from reservoirs, and irrigation water demands. The data were primarily collected from the Ministry of Water Resources and Electricity of Sudan, ENTRO's

Eastern Nile Simulation Model (ENSM) (W. N. M. Van der Krogt & Ogink, 2013), periodical reports published by the Ministry of Agriculture of Sudan (Ministry-of-Agriculture, 2013), Nile Valley Plan (MOI, 1979), and Roseires Heightening Report (McLellan, 1987).

Table 5.1 Eastern Nile Hydropower and Irrigation Systems Included in the Analysis

Name (Country)	River	Hydropower Capacity (MW)	Lateral Irrigation	
			Name	Irrigated Area (ha)
GERD (Ethiopia)	Blue Nile	5250	Beles	138,720
Roseires (Sudan)	Blue Nile	280	Upper Sennar	131,040
			Rahad	126,000
Sennar (Sudan)	Blue Nile	15	Gezira & Managil	880,000
			Ginaid	60,060
Jabel Aulia (Sudan)	Nile	28.8	Kenana	37,800
			Asalya	23,520
			WN Sugar	63,000
			WNProjects	214,200
TK5 (Ethiopia)	Tekeze–Atbara	300	---------	----------
Settit (Sudan)	Tekeze–Atbara	320	Upper Atbara	168,000
Khashm Elgirba (Sudan)	Tekeze–Atbara	10.6	New Halfa	168,420
Merowe (Sudan)	Main Nile	1250	Main Nile	230,706
Aswan High Dam (Egypt)	Main Nile	2100	---------	---------

5.2.2 Scenario development

Seven scenarios were investigated in this study. All scenarios considered the GERD reservoir to be fully developed and operational; the transient stage of filling the dam was not included in the analysis. The Eastern Nile system in Sudan was assumed to be constrained by the 1959 Agreement in all scenarios, which limits water withdrawals in Sudan to 18.5×10^9 m³/yr measured at the AHD. Each scenario was characterized by the criteria of water availability and management. Water management criteria here referred to cooperative and non-cooperative management of the system (two scenarios). Non-cooperative management means optimizing the system of each country to maximize its benefits. The first scenario (S1) is the status quo scenario, of which results have been taken from Chapter 4, where we used the RIBASIM river basin simulation model to simulate the existing system and management conditions of the Eastern Nile Basin. The simulated network in the (S1) represent the existing system in 2015 before the start of operating Settit Dam on Tekeze–Atbara River. Settit Dam has become operational since 2016; therefore, it is considered as an existing dam and irrigates 16,800 ha (**Table 5.1**). The second scenario (S2) represents the Eastern Nile system under cooperative management. The third scenario (S3) corresponds to non-cooperative management of the system. Each of water management scenarios was investigated under three water availability conditions, namely dry, normal and wet hydrological conditions (seven scenarios in total: three hydrologic conditions scenarios × two management scenarios and one status quo). The RIBASIM model developed in Chapter 4 is not an economic model, and therefore, partial comparison is conducted using the common parameters, such as generated energy, irrigation supply/demand ratios and evaporation losses.

5.2.3 Hydrological conditions considered

A monthly flow time series of 103 years of the Tekeze–Atbara, Blue Nile and White Nile (W. N. M. Van der Krogt & Ogink, 2013) were analysed to estimate 7-year periods of dry, normal and wet conditions. Ninety-six periods were generated from 103 years by taking every consecutive 7 years as one period (e.g., period-1 = year 1 to 7, period-2 = year 2 to 8, etc.). The average annual flow of every month in each period was compared with the average in 103 years of each river to define the dry, normal and average conditions (**Figure 5.2**). A 7-year time period was chosen to deal with the multi-year storage capacity of the system. The results (**Figure 5.2**) showed that the dry, normal, and wet periods of the Blue Nile and Tekeze–Atbara River occurred in 1980–1986, 1917–1923 and 1954–1960, respectively. The White Nile followed a different pattern: the dry, normal and wet periods occurred in 1920–1926, 1910–1916 and 1963–1969, respectively. Since most of Nile water is generated in the Blue Nile, and the major water resources developments will take place in Blue Nile, Tekeze–Atbara and Main Nile rivers, the hydrological periods corresponding to the Blue Nile and Atbara Rivers were considered in the analysis. The model was run on a monthly basis for each hydrologic condition (1980–1986, 1917–1923, and 1954–1960) to assess the sensitivity of optimal reservoir operation to hydrological variability. It is worth mentioning that the annual average of the monthly flow affects the operation of reservoirs because of the inter-annual variability of the Nile River, which is evident (Eltahir, 1996; Siam & Eltahir, 2015). The effect varies with the capacity of reservoir: the effect will be small, or there will be no effect in case of over-year-storage reservoirs and large in case of annual-storage reservoirs. We considered, however, the annual average monthly flows to identify the periods of dry, normal and wet conditions, because most small reservoirs in the system are controlled by the two large over-year storages as they are positioned between the upstream GERD and the downstream AHD.

5.2.4 Model parameters and assumptions

A planning horizon of 7 years (T=84 months) was used to consider the over-year storage capacity of the Eastern Nile system. The planned infrastructure considered in this study is GERD.

The large infrastructure developments in Ethiopia are assumed to be operated mainly for hydropower. It is assumed that there would not be large irrigation developments on the main stem of the Blue Nile, only Tana-Beles irrigation scheme existing upstream GERD is considered. No predefined hydro-energy demand is assumed to estimate the hydropower benefits.

The irrigation demand varies between upstream and downstream according to the crop water requirement (CWR) which depends on the cropping pattern (crop factors Kc), and reference evapotranspiration (ETo). Crop water requirements have been estimated based on FAO data (W. N. M. Van der Krogt & Ogink, 2013). **Figure 2 in Appendix-III** shows the regional variation of ETo, indicating lower CWR in the upper basins and higher CWR in the downstream Main Nile River basin. The parameter ($\propto$) representing the supply reliability (supply /demand) is used to constrain the maximum and minimum volume of water withdrawn for irrigation, the maximum withdrawal water corresponds to supply equal to demand ($\propto=1$) while the minimum amount correspond to maximum acceptable water stress for crops, assumed here as 0.8 ($\propto=0.8$).

The net price of hydropower generation and water released for irrigation are considered as 0.08 USD/kWh and 0.05 USD/m³, respectively, and are assumed identical throughout the basin. The water value impacts the optimization decision as that more water goes where the highest return can be achieved within certain boundary conditions and constraints. The water return varies between water users and countries; therefore, an economic analysis is required to estimate the water price. Such analysis is beyond the focus of the study; therefore, the economic returns are assumed the same for all countries. Similar assumptions have earlier been made by Goor et al. (2010) and Whittington et al. (2004). These values are consistent with international experience (Goor, et al., 2010). Jeuland et al. (2017) used 0.07 and 0.1 USD/kWh for hydropower price without and with power trade between countries, respectively. In our study, the energy transmission and initial infrastructure cost are not included as part of the hydropower generation.

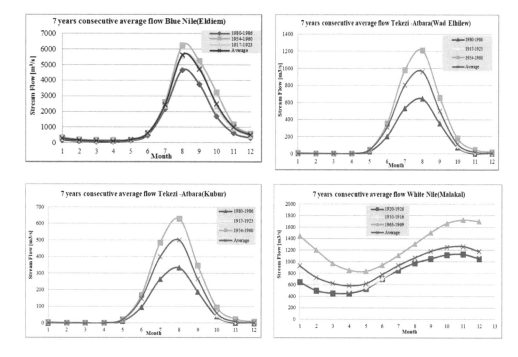

Figure 5.2. Consecutive dry, normal and wet periods with average flow of Tekeze–Atbara River, the Blue and White Nile, estimated from 103-year monthly flow data (e.g., January = month 1)

5.3 RESULTS AND DISCUSSION

The section starts with the results of the economic return from hydro-energy and irrigation of the Eastern Nile system, at the basin level, considering average (normal) hydrologic conditions, i.e., from 1917 to 1923 (water availability) and cooperation and non-cooperation between countries in managing the system. Then, a comparison of various system indicators at the country level under non-cooperative management of the system will be conducted. The section ends with the discussion on the sensitivity of the results to dry and wet hydrologic conditions. Only the sensitivity of the Ethiopian system is discussed here, because Ethiopia contributes more than 85% of the Eastern Nile water yield.

5.3.1 Cooperative versus non-cooperative system management

In the cooperative system management scenario, the Eastern Nile was optimized as one system and generates system-wide economic returns. In the non-cooperative management

73

scenario, the system within each country was optimized separately, without concern for downstream demands; releases from the optimal system state in the upstream country were used as regulated inflows for optimizing the downstream country's system. Both irrigation and hydropower objectives had the same weight, and therefore, were optimized simultaneously as a Bi-Objective optimization.

Figure 5.3 depicts the trade-off between average annual benefits of irrigation and hydro-energy at the basin level, under the two different system management conditions and under normal hydrological conditions, which were plotted for the minimum, 25th percentile, median, 75th percentile and maximum return values of hydro-energy and irrigation, taken from the population of the optimal Pareto set. The results showed that, in case of non-cooperative management, the average irrigation benefits would have a relatively wider margin (1.85 to 2.01×10^9 $/yr) compared to the hydro-energy benefits (2.91 to 2.98×10^9 $/yr), indicating the sensitivity of irrigation to the management condition. Reduction in the hydro-annual generation return by 1.0×10^6 $/yr would increase the irrigation return by 2.3×10^6 $/yr. The average annual hydro-energy benefits could increase from 2.8 to 3.1×10^9 $/yr without any change in irrigation benefits (1.95×10^9 $/yr) under the cooperative system management. The countries where irrigation is dominant would be negatively impacted by the non-cooperative management. In line with findings of Whittington et al. (2005), the results showed that the total returns collected from hydro-energy and irrigation are almost equal in both system management scenarios; however, the distribution of this return vary significantly between irrigation and hydro-energy and thus between countries. This is because the upstream country (Ethiopia) has mainly hydropower potential while the downstream countries have both hydropower (HP) and irrigation potential (Sudan and Egypt). **Table 5.2** shows the average total annual return of each country from both hydro-energy and irrigation for both management scenarios and average hydrologic conditions. The results showed that non-cooperative management would have insignificant impacts on the total annual returns for Ethiopia and Sudan, while it would reduce the total returns for Egypt by 7%. Results showed the limited negative impact of the GERD development under the non-cooperative scenario, because the GERD is a non-consumptive water user and our scenarios did not consider possible future additional water abstraction projects in Ethiopia and Sudan. The over-year storage capacity of AHD and its capability to be operated at a lower water level can further reduce these impacts. Our results support the findings of Jeuland et al. (2017), which showed that non-cooperative management would reduce the total return for Egypt by 9% compared to cooperative management.

With the GERD in place, hydropower generation would unsurprisingly increase enormously in Ethiopia (**Table 5.3**). Hydropower generation in Sudan would benefit from the presence of GERD, in both management scenarios. Interestingly, Egypt would benefit from the GERD in the cooperative management scenario, as its hydropower generation from the AHD would increase by 8.7% and 12.6% compared to the status quo and the non-cooperative scenario, respectively. The large hydro-generation capacity of AHD and its location at the most downstream of the system would encourage the system to release more water towards the AHD for maximizing the hydro-energy generation of the whole system.

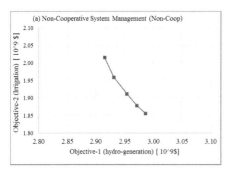 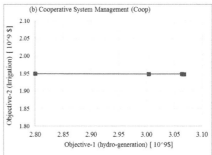

Figure 5.3. Trade-off between annual hydro-generation and irrigation benefits. Optimal Pareto Front of two objective functions over the optimization period for: (a) non-cooperative system management (Non-Coop), and (b) cooperative system management (Coop) of the Eastern Nile Basin, under normal hydrologic conditions

Table 5.2 Summary of financial returns comparing cooperative and non-cooperative management scenarios

	Ethiopia		Sudan		Egypt	
	Coop.	**Non-coop.**	**Coop.**	**Non-coop.**	**Coop.**	**Non-coop.**
Average annual returns from combined hydropower and irrigation [Million $/yr]	1,363	1,372	1,676	1,663	1,974	1,827

Table 5.3 Summary of key performance criteria comparing the status quo (without GERD) with the cooperative and non-cooperative management scenarios

	Ethiopia			Sudan			Egypt		
	Status quo	Coop	Non-coop	Status quo	Coop	Non-coop	Status quo	Coop	Non-coop
Annual energy generation [TWhr/yr]	1.38	16.4	16.8	7.6	9.55	9.39	11.5	12.5	11.1
Irrigation supply reliability (supply/demand) [%]	100	100	87.5	98.9	85.5	81.5	100	100	87.8
Annual reservoir evaporation rate [$10^9 m^3$/yr]	0.205	2.80	2.82	5.26	5.98	7.66	13.30	8.07	6.94

Irrigation supply reliability is generally sensitive to the management scenario chosen, with all three countries benefiting from cooperative management (**Table 5.3**). In this scenario, both Ethiopia and Egypt are not affected by the GERD, while Sudan sees an irrigation supply reliability decrease from 99% to 86%. This reduction is attributed to the irrigation scheme developed with Settit Dam and the presence of the trade-off between irrigation schemes and downstream hydro-demand of Merowe and AHD.

Total evaporation from reservoirs in the Eastern Nile system, in the cooperative management scenario, would decrease by about 10% with the GERD in full operation (a saving of approximately 1.9×10^9 m³/yr). The increase in evaporation from GERD would be less than compensated by a decrease in evaporation from existing reservoirs in Sudan and Egypt. Non-cooperation would increase evaporation rates in Sudan and decrease such rates in Egypt.

Figure 5.4 depicts box- and whisker-plots of monthly water levels of the GERD, Roseires and AHD reservoirs for the cooperative and non-cooperative management scenarios. The lower and upper dash lines indicate the minimum and maximum operation levels, respectively.

Typical to hydro-electric reservoirs constructed on highly seasonal rivers, the monthly water level of GERD under both cooperative and non-cooperative management scenarios

(**Figure 5.4a**) would drop (drawdown) during the dry seasons and raise (refill) during the wet season (July–October). Water levels would fluctuate more in the cooperative management scenario.

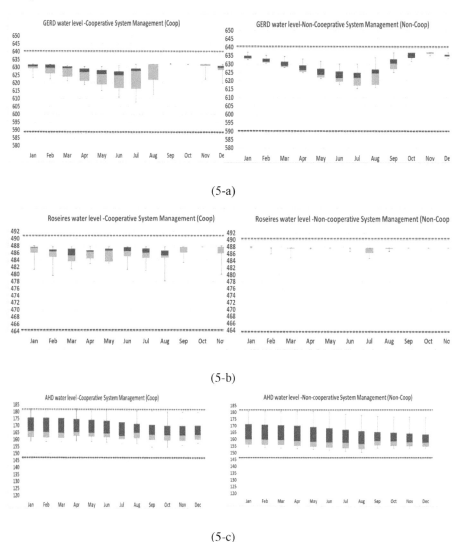

(5-a)

(5-b)

(5-c)

Figure 5.4 Boxplot of the monthly water level of GERD (a), Roseires (b), and AHD (c) for both cooperative (Coop) and non-cooperative (Non-Coop) Eastern Nile system optimization, under normal hydrological conditions

Figure 5.4b depicts water levels of Roseires reservoir. The current drawdown-refill cycle (not shown) would disappear with the GERD in place, for both cooperative and non-cooperative scenarios. Interestingly, water level fluctuations would become minimal in case of non-cooperative management.

Under the cooperative management scenario, water levels of the AHD would remain between 154 and 182 m.a.s.l., while these would reduce by 4 m when the system is managed non-cooperatively (**Figure 5.4c**). Yet, the minimum operation levels of both management scenarios would still be higher than the current minimum operation level (not shown). The drawdown-refill cycle of AHD would experience a slight shift from the normal seasonal pattern of the Nile River with the GERD in place, indicated by lower water levels in November and December.

5.3.2 Hydrologic sensitivity

The Eastern Nile was optimized for different hydrologic conditions to assess its hydrologic sensitivity. Here, we only presented the results for Ethiopia's hydrologic sensitivity for the non-cooperative management scenario, which represent the unfavourable condition for the downstream countries. **Figure 5.5** displays the edges of the optimal Pareto front of two objective functions, which are hydropower generation and irrigation of the upstream GERD for the three hydrologic conditions; dry, normal and wet. The results showed that the variation (between min and max) of energy returns is slightly higher in wet conditions. The average returns from energy varies from 1.23×10^9 \$/yr for dry, 1.33×10^9 \$/yr for normal hydrologic conditions, and 1.49×10^9 \$/yr for wet conditions, indicating that energy generation is sensitive to the hydrologic condition, as expected. The variation of the irrigation return is high under dry and normal conditions, but low under wet conditions, because there would be sufficient water to satisfy irrigation demands.

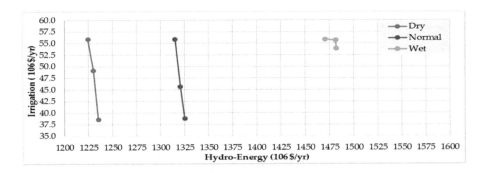

Figure 5.5 Optimal Pareto Front of two objective functions for Ethiopia's part of the Eastern Nile system for three hydrologic conditions (dry, normal and wet) for the non-cooperative management scenario

Figure 5.6 and **Table 5.4** show monthly water levels and releases for the three hydrologic conditions. The change of hydrologic conditions would not significantly change the monthly operating rules of GERD, and the minimum level is about 19 m higher than the designed minimum operation level (590 m.a.s.l.). GERD would have the capability to release the same average volume of water during dry and normal conditions (**Table 5.4**), with the minimum and maximum water releases ranging from $1.2-4 \times 10^9$ m^3/month, while under wet conditions, releases would remain constant at their maximum. The ranges of the monthly firm energy generation of GERD under dry, normal and wet conditions (**Figure 5.7**) would be 0.43–1.54, 0.58–1.57, and 1.30–1.62 TWhr/month, respectively. Compared to the average (normal) hydrologic conditions, dry conditions would reduce annual average electricity generation from 16.10 to 14.8 TWhr/yr, a reduction of 8.1%, while wet conditions would increase electricity generation by 7.9%.

Table 5.4 Monthly water levels and releases of the GERD

	Water Level (m.a.s.l.)			**Water Releases (10^9m^3/month)**		
	Dry	**Normal**	**Wet**	**Dry**	**Normal**	**Wet**
Minimum	610	615	614	1.19	1.64	4.00
Average	629	630	629	3.47	3.61	4.00
Maximum	640	640	640	4.00	4.00	4.00

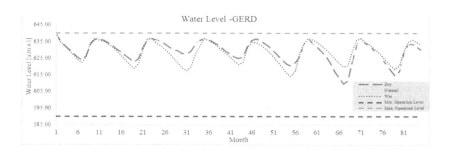

Figure 5.6 Monthly water levels of the GERD for three hydrologic conditions over the entire 7-year period considered

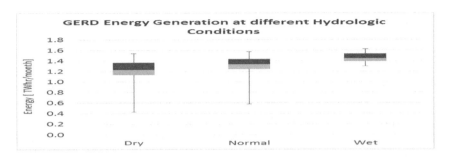

Figure 5.7 Monthly energy generation of the GERD at three hydrologic conditions

5.4 CONCLUSIONS

This study provided a quantitative analysis of the distribution of benefits resulting from the optimal operation of the Eastern Nile system, following the development of the largest hydropower generation infrastructure in the basin, the GERD. A deterministic hydro-economic optimization model for the Eastern Nile Basin, the ENOM, was developed using the GA. The analysis presented a comparison between two extreme system management scenarios, the cooperative and non-cooperative management. In the cooperative management, basin-wide system optimization was carried out, assuming full cooperation between countries to manage the whole Eastern Nile system as one entity. Non-cooperative system management considered optimizing the system within each country without taking into consideration downstream demands. Water withdrawals from the Eastern Nile system within Sudan was constrained in both management scenarios by the 1959 Nile Water Agreement. Sensitivity of the system to water availability was also analysed.

The study showed that, in case of the Eastern Nile reservoir system managed cooperatively, the basin countries could benefit from the GERD in terms of hydropower generation and maintain regulated flow, without significant change in irrigation supply. The economic return of hydropower generation and irrigation projects would be 1,363 million $/yr in Ethiopia, 9.55 million $/yr in Sudan, and 11.5 million $/yr in Egypt, compared to 1.38, 7.6 and 11.5 million $/yr, respectively, in the current situation. One worth mentioning finding is that non-cooperative management would negatively affect the irrigation sector in Ethiopia (-12.5%) and Sudan (-4%) in comparison with cooperative management; this can be explained by the geographic locations of large hydropower dams downstream of irrigation areas within these countries.

Furthermore, the results showed the sensitivity of the Eastern Nile system to changing hydrologic conditions by focusing on Ethiopia. The GERD would reduce the average monthly flow to the downstream countries under normal conditions by 13% of the historical average runoff of the Blue Nile at the location of GERD. However, the downstream countries, in particular Sudan, are hardly impacted, not even under dry conditions because of GERD's capability to regulate the flow and release almost the same volume under dry and normal conditions. Under wet conditions, the GERD would release the same volumes, on average, as the historic runoff.

The results showed the capability of the Eastern Nile Optimization Model ENOM, developed in this thesis, to optimize the Eastern Nile Basin management. The model can be used for similar basins; however, objective functions for optimization would need to be adjusted to address the basin-specific transboundary issues.

6

DEVELOPMENT OF THE EASTERN NILE RESERVOIRS SYSTEM SEDIMENTATION MODEL

Addressing sediment management in the context of optimising the operation of multi-purpose reservoirs is important, in particular for the EN where rivers carry large amounts of sediment. This chapter describes a new modelling approach for optimizing the operation of such a reservoir system considering the temporal and spatial variation of sediment deposition. The chapter also describes the application of the new model for calibration and verification to Roseires dam on the Blue Nile river in Sudan. The model is applied and shows that there is no trade-off between hydropower and irrigation water users and sediment management. Although the method was developed for multi-reservoir systems, it was applied for a single reservoir in this study due to the computation time demand and limited time available in this PhD study.

6.1 INTRODUCTION

Sedimentation poses a serious threat to the sustainability of reservoirs. The loss of storage capacity of existing reservoirs worldwide is estimated at 0.5–1% per year (Kummu, et al., 2010). Sedimentation has various levels of impacts, depending on location and capacity of reservoirs. Small-capacity reservoirs located in regions of high-sediment yield (active geological regions) are the most exposed to serious sedimentation problems. The effects of sedimentation on the function of reservoirs include: reducing the usable water storage volume, turbine damage, and interference with the outlets (Minear & Kondolf, 2009; Morris & Fan, 1998). Sediment deposition in reservoirs increases the risk of dam failure during earthquakes because of the extra forces imposed on the structure due to the higher density of sediment (Minear & Kondolf, 2009). Retaining sediment in reservoirs can have environmental and economic impacts on the upstream and downstream river and coast. Deposition of sediment can cause upstream backwater flooding. Reduction of sediment load carried downstream changes the morphology of rivers, which could in turn damage infrastructures and impact ecosystems. Lack of flooding causes shortage of sediment deposition in flood plains and deltas and disconnects the river from its flood plains. This results in fertility reduction of the areas adjacent to the river (Kummu, et al., 2010).

A wide range of sediment management strategies have been implemented worldwide to preserve reservoir storage, and these strategies are classified into three main categories (Kondolf, 2013): methods to regain reservoir capacity (i.e. flushing and dredging), methods to pass sediment through or around reservoirs (i.e. sluicing and turbid density current), and methods for watershed management to reduce sediment yield such as soil and stream bank ersoin control. Similar to sediment trap, implementation of sediment management strategies influences the functionality of reservoirs, which are generally used for hydropower generation, irrigation, domestic and industrial water supply, flood control, and recreation. For example, sluicing requires passing the high flow during the flood seasons, which would lead to a loss of head for hydropower generation and hinder flood control.

Not all sediment that is transported by upstream rivers is trapped in the reservoir. Trap efficiency (TE) is the ratio of sediment trapped in the reservoir to sediment inflow over a certain period of time. Many factors affect the TE of a reservoir — sediment properties, reservoir characteristics, such as volume, shape, area, and dam operations (Kummu, et al., 2010; Morris & Fan, 1998). Several methods are used to estimate the amount of sediment that is trapped in reservoirs, including the numerical morphodynamic model and empirical formula (Morris & Fan, 1998). The most popular empirical formulae for assessing the TE of reservoirs are those by Churchill (1948), Brune (1953) and Brown (1958).

Different methods are suggested in the literature to study reservoir operation considering sediment management. They are applicable to either existing or planned reservoirs and to single or multi-reservoir systems. **Appendix-V** provides a summary of the reviewed literature, classified based on region, method (i.e. optimization, simulation) and reservoir system condition (i.e. single/multiple, existing/planned reservoirs). Empirical formulae are generally used for planned reservoirs in reservoir simulation studies for which no field data can be found. For existing reservoirs, the most accurate method to estimate the TE is by measuring the change of reservoir volume by bathymetric surveys and then relating these changes to the inflow and outflow of suspended and bed sediment loads. Empirical formula are used in some existing reservoirs and are calibrated on the available data. Siyam et al. (2001) calibrated Brune's formula for Roseires Reservoir on the Blue Nile. Lewis et al. (2013) calibrated Brune's and Churchill's formulae for Burdekin Falls Dam reservoir in Australia at the time step of a day to fit with tropical dams, where the intra-annual variability of inflow is very high. Kummu et al. (2010) introduced a method to estimate the TE of existing and planned reservoirs in the Mekong Basin, using Brune's empirical formula. Minear and Kondolf (2009) developed a method to estimate reservoir sedimentation of a multiple reservoir system, part of which has measured sedimentation rate. They used Brown's empirical formula to estimate the TE.

There are two ways to consider reservoir sedimentation while optimizing the system. (i) Reservoir sedimentation can be considered through simulating its impact on the change of available storage for operation when optimizing water allocation. In this case, methods similar to those used in previously mentioned reservoir simulation studies can be applied. (ii) The reservoir sedimentation problem can be addressed by assessing the optimal operation for sediment management and water allocation. In this case, a predefined sediment management option (e.g., flushing, sluicing) can be added as an objective function to the overall multi-objective optimization of the system. The Tsinghua empirical formula for flushing simulation is widely used with a Genetic Algorithm (GA) to optimize an existing single reservoir (e.g., Taiwan: Chang et al., (2003); Iran: Hajiabadi & Zarghami, (2014); Ecuador: González Iñiguez, (2017)) and multiple reservoir systems (Pakistan: (Rashid, et al., 2015)). Studies of optimal sediment management of existing reservoirs can be found in the literature, but less attention is given to considering sediment management in the operation of planned dams during the design phase (Minear & Kondolf, 2009).

Reservoir sedimentation is the foremost problem of existing reservoirs in the Eastern Nile Basin (e.g. Omer et al., 2015). Loss of upstream land due to high soil erosion has resulted in an increase in sediment loads to rivers. Roseires and Sennar Dams on the Blue Nile River and Khashm Elgirba Dam on the Tekeze-Atbara River have already lost 50%, 34%, and 43% of their storage capacity, respectively, due to sedimentation (ENTRO, 2007; Gismalla, 2009). Sluicing and flushing practices are adopted in the operation of Roseires and Khashm Elgirba Dams. However, these dams still encounter reduction in their storage capacity. Unlike sluicing, flushing is considered a cost-ineffective practice for sediment management at Roseires Reservoir, given the small capacity of the reservoir with respect to the flow of the Blue Nile River, with an average flow of 50 x 10^9 m³/year, most of it occurring over three months, hand in hand with maximal sediment load. The reservoir is kept at the minimum level, with all the gates open during the flood season to pass the sediment peak, the so-called sediment sluicing. Sluicing sediment management strategy conflicts with the main objective of storing water for irrigation and hydropower generation. Large water releases through main sluices and low reservoir level reduce the benefits of irrigation, hydropower, or both. The average sediment load of the Blue Nile at Eldiem Station, near the Sudanese–Ethiopian border, is estimated at 140 million tons/year, accounted as 15% bed load and 85% suspended sediment (Ali, 2014).

In this study, a new model is developed to optimize the operation of the Eastern Nile multi-objective multi-reservoir system for hydropower and irrigation, with sediment management (**Research objective 3**). The model uses the TE concept for reservoir sediment management simulation of a multiple reservoir system. The method was first introduced by Minear and Kondolf (2009). An optimization-simulation model using GA developed in Chapter 5 (Digna et al., (2018) is applied to optimize water allocation for hydropower, irrigation, and sediment management.

This study is thus distinguished by considering the reservoir sedimentation issue in water quantity optimization of planned and existing multi-objective reservoir systems for hydropower and irrigation, which has so far been neglected in previous studies on the Eastern Nile basin. This study also investigates the temporal and spatial variation of the sedimentation rate of the system.

6.2 METHODOLOGY

In this study, the dynamic interrelation between reservoir operation and sedimentation problems is considered in optimizing reservoir operation. The available storage for optimizing the operation varies with time, depending on the decisions of operation. A reservoir system optimization and simulation with sediment management (RSOSSM) model was developed to optimize the operation of a multi-objective multi-reservoir system considering sediment management. The conceptual framework is shown in **Figure 6.1**. The model includes three modules: the optimization module, the reservoir operation simulation module, and the sediment management simulation module. The optimization and simulation model developed in Chapter 5 was adapted to include a new module for sediment management simulation. The model was coded in MATLAB 2015, and optimization was carried out using the GA available in the optimization tool of MATLAB. The optimization module assesses the objective functions given by the decision variable and reservoir operation–related parameters. The reservoirs' state of storage and water levels are estimated in the reservoir operation simulation module. Sediment deposition, TE, and updates for the reservoir Level-Area-Volume relationship are calculated from the sediment management simulation module, as illustrated in the following Section. The model runs at a monthly time step. It is developed for the Eastern Nile system and applied to the Roseires single reservoir for model calibration and verification. The model, however, can be applied to other similar systems.

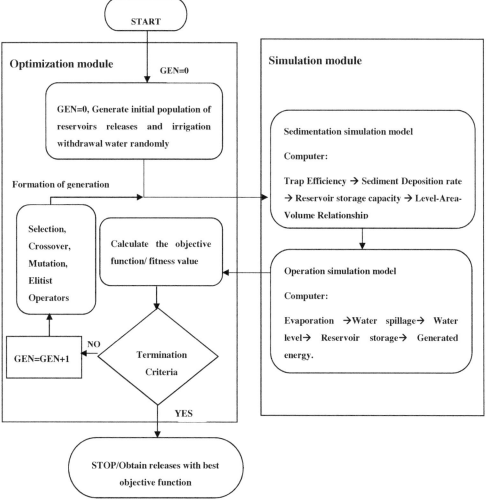

Figure 6.1 Conceptual framework of reservoir system optimization-simulation with sediment management

6.2.1 Reservoir system sediment management simulation model

Despite the sediment data limitation in the basin, few measurement data and estimates at downstream reservoirs in Sudan are documented in the literature. The following method describes a modification to the method introduced by Minear and Kondolf (2009) to compute sediment inflow to a reservoir. The main concept of the method of Minear and Kondolf (2009) is based on the following:

i. The catchment area of the tributary has homogeneous geomorphic characteristics (similarity in climate, relief, geology, and vegetation). The sedimentation yield of the tributary is assumed to be linearly proportional to the drainage area contributing to the tributary runoff. It is also assumed that the tributary is morphologically stable; there is no in-stream sediment yield resulting from bank erosion. In the case of the Eastern Nile Basin, the Ethiopian and Eritrean plateau is the source of most of the sediment yielded from the Blue Nile and Tekeze Rivers, respectively. Despite the change of geomorphic characteristics of the upstream and downstream part of the catchment, the assumption is valid because all reservoirs (new or planned) of unmeasured sediment yield data are located within the Ethiopian Highlands of similar characteristics, and the reservoirs for which measured data are available are located downstream, where geomorphic changes occur.

ii. The sedimentation rate in new (planned) reservoirs can be estimated using temporally varied TE and considering the special distribution of reservoirs (i.e., upstream reservoirs) in the same tributary.

Four functions are performed in the sediment management simulation model to estimate: (i) sediment yield rates for multiple reservoir system, (ii) trap efficiency, (iii) reservoir sedimentation rates, and (iv) reservoir new capacity and update of the Level-Area-Volume relationship.

<u>**Estimation of sediment yield rate for multiple reservoirs system [$Y_{t,i}$] (tons/month)**</u>

For a cascade of reservoirs located at the same tributary and in the part of the catchment of similar geomorphic characteristics (**Figure 6.2**), given the sediment yield at the reservoir (c) at the most downstream locations ($Y_{t,c}$) and the drainage area (Ac), the sediment yield rate at a planned reservoir, with no data located upstream of the reservoir (c), can be estimated as:

$$Y_{t,b} = \frac{A_b}{A_c} * Y_{t,c} \qquad\qquad (6.1)$$

$$Y_{t,a} = \frac{A_a}{A_c} * Y_{t,c} \qquad\qquad (6.2)$$

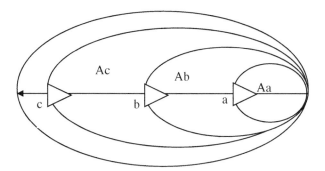

Figure 6.2 Cascade of reservoirs in the same tributary

For a cascade of reservoirs, given the sediment yield at the reservoir (a) at the most upstream locations $(Y_{t,a})$, the sediment yiled rate at a planned reservoir, with no data located downstream of the reservoir (c), can be estimated as:

$$Y_{t,b} = \frac{A_b}{A_a} * [Y_{t,a}(1 - TE_{t,a})] \qquad (6.3)$$

$$Y_{t,c} = \frac{A_c}{A_b} * Y_{t,b}[(1 - TE_{t,b})] \qquad (6.4)$$

where $Y_{t,a}$, $Y_{t,b}$, $Y_{t,c}$ (tons/month) are sediment yield rates at reservoirs a, b, and c; A_a, A_b, A_c, are the drainage areas of reservoirs a, b, and c; and $TE_{t,a}$ and $TE_{t,b}$ are the TE of reservoirs a and b.

The above conditions assume that each reservoir receives lateral flows and sediment load from its drainage area. In case the drainage area is the same for two reservoirs, i.e., $A_a=A_b=A_c$, given the sediment yiled at reservoir (a) at the most upstream locations $(Y_{t,a})$, the sediment yiled at a planned reservoir, with no data located downstream of the reservoir (c), can be estimated as:

$$Y_{t,b} = [Y_{t,a}(1 - TE_{t,a})] \qquad (6.5)$$

$$Y_{t,c} = Y_{t,b}[(1 - TE_{t,b})] \qquad (6.6)$$

Estimation of Trap Efficiency (TEt, j) (decimal)

Given the objectives of this study, the TE approach is found to be an appropriate method to simulate reservoir sedimentation, since it is simple and related to operational problems. Also, the literature shows a considerable application of the TE model in similar situations (Garg & Jothiprakash, (2009); Kummu et al., (2010); Minear & Kondolf, (2009); Mohamed, (1990)). Mohamed's (1990) concept for TE estimation is selected over other TE methods for several reasons. First, TE is expressed as a function of the optimization decision variables, releases from reservoir and storage, and thus water level that has direct impact on hydropower generation. Explicit relation between TE and optimization decision variables would relatively reduce the uncertainty associated with simplifying complex functions of sediment dynamics. Second, TE can be estimated at different time scales, such as ten days, monthly, and annually (1990). Ali et al., 2014 demonstrated that Brune and Churchill formulae overestimate the trap efficiency of Roseires dam compared to the observed values.

The trap efficiency ($TE_{t, j}$) of reservoir (j) at time step (t) is proportional to storage-capacity ratio ($STR_{t, j}$) and inversely proportional to the flushing operation ($FIR_{t,j}$). Storage- capacity ratio is the ratio between reservoir (j) storage at time t ($S_{t, j}$) and the maximum storage capacity at time t-1 ($S_{t-1,j}^{max}$). Flushing operation can be expressed as ratio between outflow ($Qo_{t, j}$) and inflow ($Qi_{t, j}$). Inflow ($Qi_{t, j}$) refers to the total unregulated lateral flow and releases from upstream reservoirs.

$$TE_{t,j} \propto \frac{STR_{t,j}}{FIR_{t,j}} \tag{6.7}$$

$$STR_{t,j} = \frac{S_{t,j}}{S_{t-1,j}^{max}} \tag{6.8}$$

$$FIR_{t,j} = \frac{Qo_{t,j}}{Qi_{t,j}} \tag{6.9}$$

$$TE_{t,j} = Csd * \frac{S_{t,j}}{S_{t-1,j}^{max}} * \frac{Qi_{t,j}}{Qo_{t,j}} \tag{6.10}$$

Where: Csd =constant (0<cs<1), can be determined from model calibration for reservoir having data. Different values for Csd between 0-1 can be tested for reservoir having no data.

Estimation of reservoir sedimentation rates (on monthly time step) [SDRt,j] (m³/month)

Sediment deposition of a cascade of reservoirs in one stream (**Figure 6.2**) can be estimated as follows:

$$SDR_{t,c} = TE_{t-1,c} * [(Y_{t,c}/\varphi) - SDR_{t,b})] \tag{6.11}$$

$$SDR_{t,b} = TE_{t-1,b} * [(Y_{t,b}/\varphi) - SDR_{t,a}] \tag{6.12}$$

$$SDR_{t,a} = \frac{TE_{t,a} * Y_{t,a}}{\varphi} \tag{6.113}$$

where: φ is the sediment dry bulk density (estimated as 1.2 ton/m³ (Ali, 2014)).

Estimation of reservoir new capacity and maxSt, j (m³) and update the Level-Area-Volume:

Sediment deposition cumulatively reduces the reservoir capacity for water storage. The distribution of sediment deposited in the reservoir results in change of Level-Area-Volume relation. The storage capacity of reservoir ($S_{t,j}^{max}$j) at time (t) after sediment deposition during time (t-1) can be estimated as:

$$S_{t,j}^{max} = S_{t-1,j}^{max} - SDR_{t,j} \qquad (6.14)$$

where $S_{t-1,j}^{max}$ is the storage capacity at time (t-1)

There are many techniques proposed to estimate the sediment distribution within the reservoir, such as the Empirical Area Incremental method, and the Area Reduction method (Ali, 2014; Rashid, et al., 2015). A simple and practical concept is applied to update the Level-Area-Volume relationship, since the detailed distribution of sediment with reservoirs is beyond the scope of this study. The Level-Area-Volume relationship is calculated assuming the intercept in the original relationship remains constant, and only the slope varies with the water level, since sediment deposition is mostly anticipated to deposit upstream along the reservoir and less would be deposited in front of the gate because of sluicing. The Level-Area method is used to calculate the reservoir surface area given water level at each time step to estimate the evaporation loss from the reservoir. The Level-Volume relationship is used to identify water levels (head) corresponding to the storage at each time step, which is used to estimate the hydropower generation as described in Section 6.2.2, below.

6.2.2 Optimization model

The optimization problem is formulated to maximize the aggregated net benefits associated with water allocation for hydropower generation (f1) and irrigated agriculture (f2) and minimize the sediment deposition (f3) by identifying optimal turbine release, irrigation withdrawal, and release for sediment flushing (R_t) at each time step (t) over time horizon (T). The problem is mathematically formulated as follows (see also Chapter 5):

Objective function

Three objective functions are developed to maximize the returns from:

    (i) hydropower generation (f1)),
    (ii) irrigation releases (f2) and
    (iii)sediment released (f3)

iii. of the system during the time horizon (T). Minimizing sediment deposition is converted to maximization by multiplying the accumulated sediment by (-1) in the objective function. The overall objective function (F) is written as:

$$F(S_t, I_t, Rw_t, Rs_t, IR_t) = \max_{R_t}\{\omega_1 f1(\cdot) + \omega_2 f2(.) + \omega_3 f3(.)\} \qquad (6.15)$$

$$f1 = P_e \sum_{t,j}^{T,J} HP_{t,j} \qquad (6.16)$$

$$HP_{t,j} = c * \tau_{t,j} * \eta_{t,j} * H_{t,j}^{net} * R_{t,j} \qquad (6.17)$$

$$f2 = P_w \sum_{t,i}^{T,I} IR_{t,i} \qquad (6.18)$$

$$f3 = P_s \sum_{t,j}^{T,J} SR_{t,j} \qquad (6.19)$$

where $SR_{t,j}$ is the sediment released from reseroivr (j) at time (t), $\omega_1, \omega_2, \omega_3$ are weight factors of the respective objective functions and satisfy the condition $\omega_1 + \omega_2 + \omega_3 = 1$, Rt is the vector of decision variables that represent the reservoir's releases through the turbines ($Rw_{t,j}$), releases for irrigation ($IR_{t,i}$), and releases for sediment flushing ($Rs_{t,j}$) at each time step (t). The vector is in the following form:

$[R_t] = [Rw_{1,1} \ Rw_{2,1},.., Rw_{T,1}; \ Rw_{1j} \ Rw_{2,j},.., Rw_{T,j};..; Rw_{1,J} \ Rw_{2,J},.., Rw_{T,J}; IR_{1,1} \ IR_{2,1},.., IR_{T,1};$
$IR_{1,i} \ IR_{2,i},.., \ IR_{T,i};..; IR_{1,I} \ IR_{2,I},.., \ IR_{T,I} ; \ Rs_{1,1} \ Rs_{2,1},.., Rs_{T,1}; \ Rs_{1,j} \ Rs_{2,j},.., Rs_{T,j};..; Rs_{1,J} \ Rs_{2,J},..,$
$Rs_{T,J};]$

Releases for the three water users are discretized in optimization model to allow allocating water for water users of different priorities and scenarios development. These releases are however used conjunctively for the three users per each user constraints similar to the real operation of the reservoir.

Constraints

The objective function is subject to the following constraints:

Energy generation constraints:

$$HP_{t,j} \leq HP_{t,j}^{max} \qquad (6.20)$$

$$q_{t,j}^{min} \leq Rw_{t,j} \leq q_{t,j}^{max} \qquad (6.21)$$

Reservoir storage limits:

$$S_j^{min} \leq S_{(t,j)} \leq maxS_{(t,j)} \qquad (6.22)$$

where $maxS_{(t,j)} = S_j^{max}$ at the initial condition $(t = 1)$

Irrigation withdrawal limits:

$$\alpha * (A_i * CW_{t,i}) \geq IR_{t,i} \leq (A_i * CW_{t,i}) \tag{6.23}$$

Continuity (mass conservation) constraint:

$$S_{t+1,j} = S_{t,j} + I_{t,j} + C_{j,k}^R (Rw_{t,j} + Rs_{t,j} + Sp_{t,j}) + C_{j,z}^{IR} (IR_{t,i}) - e_{t,j} \tag{6.24}$$

$$e_{t,j} = A_{oj} * Ev_{t,j} + A_{tj} * Ev_{t,j} * (S_{t+1,j} + S_{t,j})/2 \tag{6.25}$$

$$Sp_{t,j} = S_{t+1,j} - maxS_{(t,j)} \quad \text{if } S_{t+1,j} > maxS_{(t,j)} \quad \text{Otherwise,} \tag{6.26}$$

$$Sp_{t,j} = 0$$

End storage constraint:

$$\forall_j, S_{T,j} \geq D_j \tag{6.27}$$

Non-negativity constraints:

$$Rw_{t,j}, Rs_{t,j}, S_{t,j}, IR_{t,i}, HP_{t,j}, TE_{t,j} \geq 0 \tag{6.28}$$

Required flow downstream Roseires dam:

$$DSD_{t,j} \leq Qo_{t,j} \leq Qo_{t,j}^{max}, \quad Qo_{t,j} = Rw_{t,j} + Rs_{t,j} + Sp_{t,j}, \quad \forall_{t,j} \tag{6.29}$$

Where:

Symbol	Unit	Description
$HP_{t,j}$	MWh/month	Total generated energy from reservoir (j) at time (t)
Pe	USD/MWh	The economic benefit of generated energy
c	N/m^3	Constant represents specific gravity and unit conversion
$\tau_{t,j}$	hours/month	Number of hours in period (t)
$\eta_{t,j}$	-	Turbine efficiency
$H_{t,j}^{net}$	M	Turbine net head of reservoir (j) at time (t)
$Rw_{t,j}$	m^3/month	Turbine release from reservoir (j) at time (t)
P_w	USD/m^3	The economic benefit of withdrawal water for irrigation
$IR_{t,i}$	m^3/month	Withdrawn water for irrigation (i) at time (t)
$Rs_{t,j}$	m^3/month	Sediment flushing release from reservoir (j) at time (t)
$TE_{t,j}$	-	Trap efficiency of reservoir (j) at time (t)
$Rs_{t,j}$	m^3/month	Water for Sediment flushing of reservoir (j) at time (t)
Ps	USD/ m^3	The economic benefit of released sediment

93

Symbol	Unit	Description
$HP_{t,j}^{max}$	MWh/month	Maximum hydropower energy generated from reservoir (j) at time (t)
$q_{t,j}^{min}$	m³/month	Minimum turbine discharge of reservoir (j) at time (t)
$q_{t,j}^{max}$	m³/month	Maximum turbine discharge of reservoir (j) at time (t)
$S_{t,j}$	m³/month	Storage state variable of reservoir (j) at time (t)
S_j^{min}	m³	Minimum storage volume of reservoir (j)
S_j^{max}	m³	Maximum design storage volume of reservoir (j)
D_j	m³	Target end storage of reservoir (j) at time (T)
DSD_j	m³/month	The minimum flow required downstream Roseires Reservoir, including environmental flow and downstream demand at Khartoum, estimated at 244 x 106 m3/month
$Qo_{t,j}^{max}$	m³ /month	Maximum release capacity of the reservoir at time (t)
$Ev_{t,j}$	m/month	Monthly Evaporation rate from unit area of surface of reservoir (j) at time (t)
A_i	m²	Irrigated area of scheme (i)
$CW_{t,i}$	m/month	Crop water requirement of irrigation scheme (i) at time (t)
$\propto$	-	Coefficient representing supply/demand ratio
Ev_t	m/month	Monthly Evaporation rate from unit area of surface
$I_{t,j}$	m³/month	Inflow state variables at reservoir site (j) at time (t)
$Sp_{t,j}$	m³/month	Spillage of reservoir (j) at time (t)
$e_{t,j}$	m³/month	Evaporation loss of reservoir (j) at time (t)
A_{oj}	m²	Surface area of reservoir (j) at dead storage level
A_{tj}	m²/m³	Area per unit storage of reservoir (j)
$C_{j,k}^R$	-	Reservoir system connectivity matrix=-1 when abstraction, +1 receive water from upstream reservoir [reservoir (j) receives water from reservoir (K)]
$C_{j,z}^{IR}$	-	Irrigation system connectivity matrix= -1 when abstraction, +1 receive return water from upstream irrigation [reservoir (j) receives water from irrigation (i)]
T	Month	Planning time horizon
J	-	Total number of dams in the system
I	-	Total number of irrigation schemes in the system

Further discussion on the assumptions used to define constraints is given in section 6.3.3.

6.3 STUDY AND SCENARIO DEVELOPMENT

6.3.1 Case study

Roseires Dam is selected as the case study for model calibration and verification, for several reasons. First, Roseires Dam is located at the Blue Nile River, which contributes to more than 80% of the Nile total sediment load, which is estimated at 160 million tons/year at Aswan High Dam (AHD) (Ahmed & Ismail, 2008). Second, the location of

Roseires Dam can be considered a divide for the Blue Nile catchment on the geomorphologic bases into upstream, the source of sediment and water runoff, and downstream that conveys water and sediment load without significant contribution. Third, Roseires Dam is the first sediment trap in the Blue Nile (Ali, 2014) (**Figure 6.3**). Fourth, despite sediment data limitation in general in the basin, there are relatively fair sediment data that can be used for model verification.

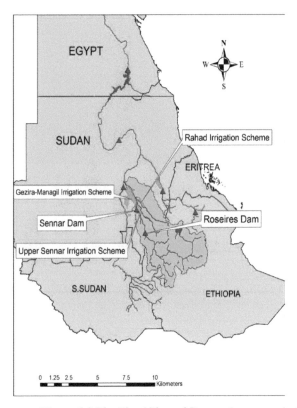

Figure 6.3 The Blue Nile and Reservoirs system in Sudan

Roseires Dam is a multi-objective dam for hydropower generation and irrigation that started operation in 1966. The dam was developed to support the operation of Sennar Dam, to satisfy the downstream demands. The characteristics of the river and the dam are shown in **Table 6.1**. The reservoir lost more than 50% of its capacity (**Table 6.2**) because of sediment load carried by the river, estimated at 140 million tons/year. **Figure 6.4** displays the average monthly inflows and the percentage distribution of sediment flow of the Blue Nile at the location of Roseires Dam. The operation of Roseires was changed a few years after commissioning, as a result of sediment deposition, so that filling starts after the peak sediment load pass. This is generally anticipated between the 1st and 26th of September, annually, depending on the amount of water inflow, and continues for forty-five days. The dam is kept at the minimum operation level of 467 during the first two months of the flood period, as shown in the operation rule of Roseires Dam in **Figure 6.5**. After the filling period, abstraction from the reservoir continues from November to April, depending on the downstream irrigation demands. The reservoir is operated with priority given to irrigation over hydropower demands.

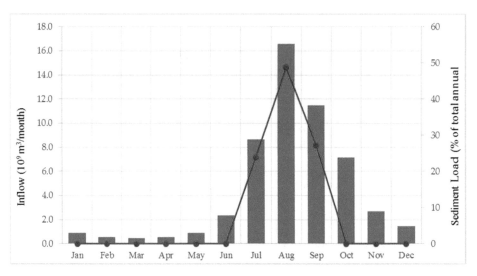

Figure 6.4 Average monthly inflow and percentage of sediment inflow of the Blue Nile at the location of Roseires dam

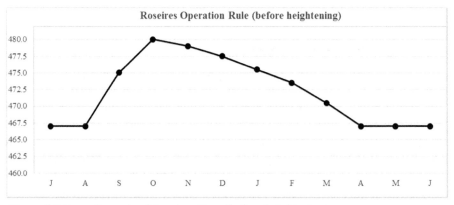

Figure 6.5 Operation rule of Roseires Dam (before heightening)

Table 6.1 Summary of Roseires Dam characteristics

Roseires Reservoir before Heightening

Completion year	1966
Catchment area [km^2]	177.2 x 10^3
Annual runoff [109 m^3/year]	50
Evaporation loss [10^9 m^3/year]	0.405
Total sediment load (including bed load) [10^9 kg/year]	140
Design reservoir length (before heightening) [km]	75
Design reservoir area (before heightening) at level 480 m [km^2]	290
Maximum operation level (m.a.s.l.)	481
Minimum operation level (m.a.s.l.)	467.0
Design storage capacity at maximum operation level [10^9 m^3]	3.024
Live storage at full supply level [10^9 m^3] (1992)	2.020
Number of spillways	10
Number of sluice-gates	5

Table 6.2 Trap efficiency, storage capacity and deposited sediment of Roseires Reservoir (Ali, 2014; A. M. Siyam et al., 2005)

Years	Operation years since 1966	Trap efficiency (%)	Storage capacity (10^6 m^3) at Level 481	Deposited Sediment (10^6 m^3)
1976	10	45	N/A	550
1981	15	36	N/A	665
1985	19	33.2	2337.6	1102
1992	26	28	2191.6	1225
2005	39	26.1	N/A	1394.3
2007	41	24	1953.8	1408.1

6.3.2 Scenario development

Three scenarios of operation policies are studied by giving a variety of priorities for the main objectives of reservoir operation (hydropower generation and irrigation) and sediment management, described below. Similar to the real operation of Roseires dam, irrigation demands are given priority over hydropower generation, by giving hydropower half the weight of irrigation in the objective function (w1=1/3, w2=2*w1). The total sum of weight is equal to one.

(i) The first scenario (S1) maximizes the weighted economic return of hydropower generation and irrigation withdrawal, irrespective of reservoir sedimentation. Here, reservoir sedimentation is incorporated in the model, but no sediment flushing is performed (i.e. Rst is set to zero as shown in **Table 6.3**). Sediment load carried by

97

water releases through turbines and for irrigation is used to estimate the trap efficiency and the economic return of sediment released.

(ii) The second scenario (S2) prioritizes sediment management over hydropower generation and irrigation in the objective function during the flood season (June to September), when the reservoir receives most sediment load from upstream rivers (Table 6). In this case, a decision variable for additional water release for sediment flushing (Rst) is optimized during flood season, with maximum release being equal to the maximum release capacity of the dam. The upper and lower limits of decision variables of irrigation and hydropower are assumed to be zero during the flood generation. Hydropower generation and irrigation are yet optimized during the remaining periods (**Table 6.3**). Although the hydropower generation and irrigation are not included in the objective function during the flood season, water releases for flushing are used to fulfil irrigation demand and generate hydropower.

(iii) The third scenario (S3) is designed to maximize the economic return of hydropower generation, irrigation, and sediment management, with equal priority weight given to the latter two (**Table 6.3**). Decision variables for hydropower, irrigation and sediment flushing are optimized with upper bounds set as the maximum generation capacity, maximum irrigation demand and maximum dam release capacity for sediment management, respectively.

Table 6.3 Priority weights (w) for each objective function at the three scenarios

Scenario	S1		S2		S2	
Objective weights	October -May	June- September	October -May	June- September	October -May	June- September
Hydropower (w1)	0.33	0.33	0.33	0	0.33	0.2
Irrigation (w2)	0.67	0.67	0.67	0	0.67	0.4
Sediment management (w3)	0	0	0	1	0	0.4

6.3.3 Model parameters and assumptions for the case study

The optimization model parameters are shown in **Table 6.4**. The economic value of hydropower generation and water released for irrigation are assumed at 0.08 USD/kWh and 0.05 USD/m3, respectively. These values are consistent with international experience (Goor, et al., 2010; Jeuland, et al., 2017; Whittington, et al., 2005). The economic value of released sediment (Ps) is identified based on the "cost avoided" concept. The unit cost of new dam storage can be avoided when a sediment management policy is adopted in the operation to reduce sediment deposition and loss of water storage capacity. Construction cost of some storage dams in the region are used as indicator for avoided cost, shown in **Table 6.5.** The unit cost of storage of the recent Roseires heightening, estimated at 0.33 $/m^3, is used to estimate the economic return of sediment management. Heightening of Roseires dam could have been avoided if the sediment deposition had not significantly reduced the usable storage. The sensitivity of operation parameters to this adopted return of sediment management is assessed.

Irrigation demand is estimated based on fixed irrigation area and cropping pattern. Chapter 5 provides details on irrigation demand. Gezira and Managil irrigation schemes (880,000 ha) as well as Upper Sennar irrigations scheme (131,000 ha) and Rahad irrigation scheme (126,000 ha). The total release is constrained by the minimum downstream water requirement including environmental flows, estimated at 8 x 10^6 m^3/day (244 x 10^6 m^3/month).

The system is optimized for a twenty-year time horizon, from 1993 to 2012. The average monthly sediment load percent and inflow of optimization period is shown in **Figure 6.4**. Despite the environmental changes resulted from urbanization and human activities upstream, sediment load over the 7 years (1993-1999) are assumed to be the same over the 20-years of optimization period, due to data limitation. The TE coefficient (Csd) is calibrated by performing several runs for the simulation model using different values of Csd (**Table 6.6**), against known TE in 1985 and 1992 from survey measurements, as well as actual storage (St), actual storage capacity (S_t^{max}), water outflow (Qot), inflow (Qit) and sediment load (Yt) during the period 1985–1992. The calculated TE by Csd value of 0.403 (**Table 6.6**) is approximately equal to that observed in 1992 (**Table 6.2**). The calibrated value of Csd is small compared to the value 0.78, calibrated by Mohamed (1990) for sediment deposited during 1966–1983 and using a time step of ten days. The variation of Csd value is attributed to the difference in the period and time step used for calibration; this is explained by the fact that the TE reduces as more sediment accumulates in the reservoirs. Csd is calibrated in this study for the period of 1966–1992, using a monthly time step for estimating TE, and assumed constant. **Table 6.6** shows the calculated TE for different values of Csd.

Table 6.4 Optimization model parameters used in the study

Parameter	Value
Economic value of hydropower generation [USD/kWh]	0.08
Economic value of water withdrawn for irrigation [USD/m^3]	0.05
GA-Population size	1000
GA-Number of generations	5000

Parameter	Value
GA-Selection methods	Tournament method and size=2
GA-Crossover fraction	0.8
GA-Crossover function	Intermediate
GA-Mutation fraction	0.025
GA-Mutation function	Constraint dependent
GA-Stopping Criteria-Objective Function Tolerance	1×10^{-6}
GA-Stopping Criteria-Constraints Tolerance	1×10^{-5}

Table 6.5 Construction cost of some dams in the region

	Merowe	Roseires Heightening	Upper Atbara Complex Dam	GERD
Total Cost [10^9 USD $]	2.4	1	1.4	6.6
Storage [10^9 m^3]	9	3	3.5	74
Cost per storage unit [$/m^3]	0.27	0.33	0.40	0.09

Table 6.6 Calibration of trap efficiency coefficient

Trial	Csd	The calculated Trap Efficiency (TE)-1992
1	1	0.696
2	0.2	0.139
3	0.3	0.209
4	0.4	0.278
5	0.41	0.285
6	0.403	0.280

6.4 RESULTS AND DISCUSSION

This section presents and discusses the results of the three different scenarios of optimized operation of Roseires Dam, including sediment management, and compares them with current operational practice. GA solutions showed improving values of the objective function with the evolution of generations; the calculations were terminated when the difference between objective function values was less than 10^{-6}.

Table 6.7 presents the economic return and trap efficiency (TE) of Roseires Dam resulting from the three scenarios of optimization. The results showed that the average annual economic return of hydropower generation, irrigation and sediment management would be 136, 549 and 77 million USD, respectively, when sediment management is considered in the optimization (S2). Returns to irrigation and hydropower generation would reduce by 6% when priority is given to irrigation and hydropower generation during the flood season (S1), resulting from sediment deposition and loss of storage. Sediment accumulation is the highest in S1 compared to scenarios S2 and S3 because water is released only for irrigation and hydropower generation based on the upper and lower boundary of each, and more water storage would occur and thus water head would be created in favor of hydropower generation. Since there is no possibility for additional releases during the flood season compared to S2 and S3, hydropower generation in S1 occurred by the head created from increased storage, therefore more sediment has been deposited. Despite the lower water head availability in S2, the hydropower is generated by discharge as it is generated from water released for flushing which is greater than the required turbine releases for energy generation.

The current operation of Roseires is similar to S2 in giving the priority to sediment management during the flood season. The economic return would slightly decrease when the operation is optimized for all three objectives compared to S2, by 1% for hydropower generation and 4% for irrigation.

Trap efficiency (TE), the percentage of deposited sediment of the total sediment inflow, was observed at 26% in 2005 under current operation. The optimization results show that the average TE over the 20-year period is 54% when the operation focuses on irrigation and hydropower generation (S1), compared to the other two scenarios. As highlighted earlier, the current operation considers sediment management as the main objective during flood season. The results show improvement in reservoir operation in terms of sediment management. The TE of Roseires Dam in 20 years is estimated at 28% and 31% for S2 and S3, respectively (**Table 6.7**).

The average irrigation supply as percentage of the minimum crop water requirement is shown in **Table 6.7**. Despite optimizing reservoir operation for irrigation in (S1), the average monthly supply-demand ratio is estimated at 98%, with deficits occurring 10% of the time. These instances, where releases for irrigation were less than the minimum requirement, occurred mostly in the second half of the 20 year optimization period. This deficit is attributed to the reduction of useable storage resulting from sediment deposition during the 20 years. The ratio of irrigation supply to the minimum crop water requirement are 100% in the case of S2 and S3, where loss of storage is reduced by including sediment management in the operation. **Figure 6.6** illustrates the monthly water requirements from Roseires reservoir and the reservoir releases of the three optimization scenarios. The average downstream demands estimated at 870 x 10^6 m^3/month and minimum of 244 x 10^6 m^3/month, include the required environmental flows at Khartoum. The results show slight variation between the monthly releases of the three scenarios; the total annual releases, however, remain similar. The minimum monthly releases are 304, 442, and 428 x 10^6m^3/month, in S1, S2, and S3, respectively.

Figure 6.7 presents box plots of the water levels of Roseires Dam for the three optimization scenarios. The drawdown–refill cycle is shown in the three scenarios; however, water levels for S1 are kept higher compared to those in the other scenarios. Unlike in S2 and S3, the reservoir in S1 would be filled during the start of the flood season and would remain full for a relatively longer time in favor of hydropower generation. Water levels show similar patterns when considering sediment as the objective, as in S2 and S3, with a lower level in August, and filling of the reservoir occurring in September, after passing the sediment peak flow that generally occurs in August. The results show that optimizing irrigation and hydropower considering sediment management (S3) would incur a drop in water level compared to second scenario (S2) in the pre- flood season (January–May) and post flood season (October–November).

Fluctuations are shown in the optimized operation rules when sediment management is considered, as in scenario (S2) and (S3), compared to the current smoothed operation rules. The fluctuation is related to the stochastic characteristics of GA that modify releases to maximize the objective functions. The optimized operation rule can be smoothed by restricting the variation of some parameters, such as change of water levels; however, this restriction might reduce the maximum return of the objective functions. Therefore, operation rules were smoothed without changing the objective function values.

Table 6.7 Summary of the annual economic return (million USD/year), and trap efficiency of the optimization scenarios

Scen ario	Economic Return (Million USD/year)				Trap Efficiency (for 20 years)	Storage loss (in 20 years)	Irrigation Supply/ Demand ratio [1]	
	Hydro power	Irrigation	Sedime nt	Total	TE (%)	%	%	Frequency
S1	128	520	38	686	54	78.7	98%	216/240
S2	136	549	77	762	28	41.1	100%	240/240
S3	131	543	75	748	31	44.6	100%	240/240

[1.] Average monthly irrigation supply is calculated as the percentage of the minimum crop water requirement, assumed as 80% of total crop water requirement.

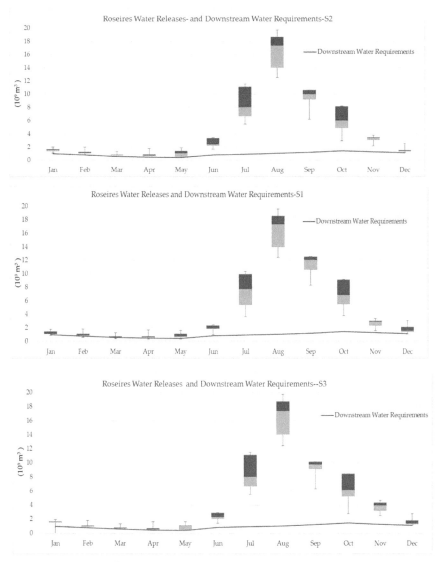

Figure 6.6 Releases from optimization scenarios and monthly downstream water requirements (including Gezira and Managil irrigation demand), averaged for the period of 1993–2012

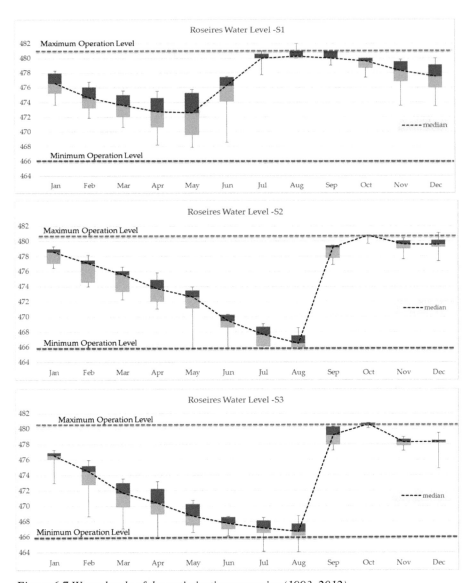

Figure 6.7 Water levels of the optimization scenarios (1993–2012)

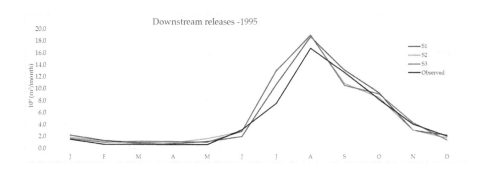

Figure 6.8 Downstream releases of Roseires Dam from existing and optimized operation rules in 1995

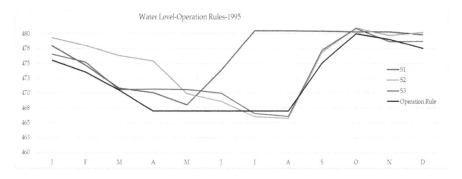

Figure 6.9 Reservoir water levels of the optimization scenarios and current operation rule in 1995

Table 6.8 Summary of sediment inflow, outflow, deposited, and trap efficiency from optimized scenarios in 1995, and observation

Scenario	Inflow	Outflow	Trap Efficiency	Deposited
	10^6 (m³/yr)	10^6 (m³/yr)	(%)	10^6 (m³/yr)
1		89	47.2	79
2	167.4	126	25.1	42
3		119	29.3	49
Observed		101.36	39.5	66.04

The releases from Roseires reservoir for the three scenarios are compared with observed releases in 1995, as shown in **Figure 6.8**. The results show that the releases from the scenarios are generally compatible with the existing releases, because the storage capacity of the dam is by far less than the inflow. More water would, however, be released when

105

sediment management is conducted in the period July–August, while less would be released in September. The increase of releases during August in S1 is attributed to the filling of the reservoir that would start in May and become full in July, while the objective of reducing sediment deposition in S2 and S3 is the reason for releasing more water during July and August. Results show no significant differences in the total releases when sediment management is considered in optimization, as in S2 and S3.

Figure 6.9 compares water levels of the three scenarios with the operation rule for 1995. The actual water level in 1995 is assumed to represent the theoretical operation rule used for operating Roseires Dam. The results show that water levels in the modelled scenarios are kept higher compared to observed levels, with the highest water levels occurring when optimizing the operation for hydropower and irrigation only (S1). Unlike in the observed operation, in S1 the filling starts as early as May and continues until July, and the reservoir is kept full until December. Drawdown starts in January and continues until April, compared to continued drawdown starting from November until August in the observed case. Taking sediment management as the objective (S2 and S3), the optimized operation would be compatible with the current operation of Roseires during the flood season (July–October), which aimed to increase the amount of sediment outflows. Results show higher water levels from December to June (S2) and April to June (S3) compared to current operation, to create head for hydropower, drop in August when the maximum sediment flows occur, and rise again to the maximum water level in October.

Table 6.8 summarizes the comparison of sediment outflow and TE computed from optimized scenarios with the measured sediment outflow in 1995. The results show that TE of the optimized scenario would be 47.1% when sediment management is not considered as an objective. The TE would drop to 25% when considering only sediment management as the objective during flood season. Optimization of operation for the three objectives — hydropower, irrigation, and sediment management — would result in a TE of 29%. The record shows that the TE resulting from the current operation is 39.46%, estimated from sediment inflow and outflow values. In case of S3, TE is higher compared to S2 because hydropower generation is included as objective function. Water level would rise in June to create head in favor of the hydropower generation, allowing deposition of incoming sediment. **Table 6.8** also shows that, when considering sediment management as the main objective during the flood season (S2), sediment deposition would reduce by 36% compared to sediment deposition observed from existing operations. In S3, the reduction would reach 25% when sediment, hydropower, and irrigation are optimized. Sediment deposition would, however, increase by 19% when operating the reservoir with the objective to maximize hydropower and irrigation only. The results demonstrate that the optimized operation rules in S2 and S3 enhance sediment management of Roseires Reservoir compared to the existing operation.

6.5 SENSITIVITY ANALYSIS

The optimization results presented above are based on an economic return equal to 0.33 USD/m^3 of sediment released. The sensitivity of the optimization framework to changes in this value is examined by how it would change water levels in the reservoir, the quantity sediment released, hydropower generated and irrigation water supplied. The sensitivity of the operation parameters is assessed by calculating the elasticities (E), which refer to the percentage change of dependent variables divided by the percentage change of independent variables (Pannell, 1997), expressed as:

$$E = (\partial Y/\partial X). \ (X/Y) \qquad\qquad (6.30)$$

Similar to "Slope", Elasticity provides quantitative measures for the rate of change of dependent variables with respect to the change of independent variables; however, elasticity overcomes the difficulty of comparing the rate of change of different parameters where the unit of measure is not comparable.

Table 6.9 presents the price elasticities of the reservoir operation parameters resulting from an increase and decrease of the economic return of sediment management by 25% and 50% of the base price (0.33 USD/m^3). The values in the table are the average monthly percentage change of operation parameters divided by absolute value of the percentage change of sediment management return. The sign (+/-) indicate the increase/reduction of parameter value resulted from the absolute change of sediment management price. The price is considered elastic or the dependent variables are sensitive to the change of price if the elasticity value is greater than 1 or less than -1.

Table 6.9 Sensitivity of selected operational parameters for the assumed cost of water storage capacity

Operation Parameters	Change of Economic Return of Sediment Management (from 0.33 USD/m^3)			
	-50%	-25%	+25%	+50%
Water level	0.000	0.000	0.000	0.000
Released sediment	-0.006	-0.002	0.004	0.007
Generated hydropower	0.020	0.013	-0.015	-0.035
Irrigation Supply	0.014	0.002	-0.007	-0.018

The results in **Table 6.9** show that (a) the elasticies have the expected sign (e.g. when the value of sediment increases, hydropower decreases) and (b) all elasticities are close to 0, and therefore not sensitive to the change of the chosen value of sediment. The largest elasticity (in absolute terms) is found with hydropower: if the value of sediment increases by 50%, hydropower generation will reduce by only 3.5%. These small changes may be because other factors may affect the changes such as the sequential relation of operation parameters over time.

6.6 CONCLUSIONS

This study presented a new methodology for optimizing the operation of a multi-objective multiple reservoir system considering sediment management. A new hydro-economic model for reservoir system optimization and simulation with sediment management (RSOSSM) was developed. The model combines GA with reservoir operation and sediment management simulation models. The optimization module assessed the objective functions given the decision variable and reservoir operation–related parameters. The reservoir's state of storage and water levels were estimated in the reservoir operation simulation module. Sediment deposition, TE, and updates for the reservoir Level-Area-Volume relationship were calculated from the sediment management simulation module. The model was calibrated and applied for an existing single, multi-objective reservoir at the Blue Nile system of Sudan, Roseires Dam. The analysis focused on assessing different levels, considering sediment management in optimization of reservoir operation, using a scenarios approach. The results were also compared with the observed data of current operation, as obtained from the literature.

The results showed that considering sediment management in reservoir operation would increase the economic return of hydropower generation (+8 million $/yr) and irrigation ((+29 million $/yr), because of utilising storage that is maintained through flushing. The study concluded that there is no trade-off between sediment management and water users, including sediment management in operation rule would not negatively impact hydropower and irrigation returns.

Despite the model and suggested method are suitable for multi-reservoir systems, this has been applied for a single reservoir because of the limited time available in this PhD study and the computation demand of complex system problems. The model is currently being run for the multiple reservoir system of the Eastern Nile multiple reservoir system, including the Grand Ethiopian Renaissance Dam; the results of which will be hopefully published in a future article.

7

CONCLUSIONS AND
RECOMMENDATIONS

The Eastern Nile river basin hosts more than 200 million people, who live in four countries, Ethiopia, South Sudan, Sudan and Egypt. In addition to the large water projects that already exist, the Eastern Nile basin witnesses increasing competition over water resources to support the growing population demands and developmental needs in those countries. This study is a scientific contribution to support decision making for optimal water resources management, both at national and basin scales. Therefore, it can be considered a contribution to support peace and sustainability in the region.

First, the study prepared a comprehensive review of the literature on water allocation modelling techniques used in the Eastern Nile basin to identify relevant knowledge gaps. Second, a river basin simulation model was developed and parameterised based on the hydrology of 103 years to study the effect of unilateral and cooperative water management of the reservoir system in the Eastern Nile. The same model was used to assess the impacts of upstream dam developments (e.g., the Grand Ethiopian Renaissance Dam) on the existing water use in the downstream countries of Sudan and Egypt. Next, the simulation model was equipped with an optimization algorithm to derive optimal operations for both national and basin scale systems. Acknowledging the importance of sediment management for the sustainable use of the reservoir system in the Eastern Nile, a new optimization model was developed. This model optimizes water allocation for hydropower generation and irrigation water supply, while dynamically considering reservoir sedimentation effects.

This chapter presents the main conclusions of the literature review in section 7.1, the river basin simulation model in section 7.2, optimal operation of the system in section 7.3, and sediment management in section 7.4. Relevant recommendations to support future sustainable development of water resources in the Eastern Nile basin are presented in the final section.

7.1 NILE RIVER BASIN MODELLING TO SUPPORT WATER RESOURCES MANAGEMENT

The review of the Nile River Basin modelling studies for planning and management concluded that some of the models use time series that are too short given the natural climate variation and their findings can therefore be misleading. Most models point at the basin-wide benefits of reservoir developments, the increased hydropower production potential and the possible expansion of irrigated agriculture enabled by these reservoirs. None of the models, however, quantified the political dimensions and societal, economic and environmental risks associated with such developments, which could possibly explain why certain developments are opposed by some riparian countries.

More than 75% of the reviewed studies simulate only parts of the Nile River Basin, the majority of which having focused on the Blue Nile sub-basin, being the largest contributor of water to Nile flows. Less attention has been given to the Tekeze-Atbara and Baro-Akobo-Sobat sub-basins. The Equatorial Lakes/White Nile sub-basin has not been studied extensively on its own but features in the studies that cover the entire Nile River Basin. Therefore, it is recommended that there is merit in carefully studying the two main sub-systems of the Nile River Basin, i.e. the Eastern Nile and the Equatorial Lakes regions, and assess how the many different development options affect the countries riparian to both sub-systems. For example, the combined impact of the planned developments on the different sub-basins in the Eastern Nile region, including the Main Nile in Sudan, has so far not been established.

Climate change is a recurring theme in most studies, but studies come to sometimes diametrically opposing conclusions. The variation on the impact of climate change is probably due to the type of the climate change scenario and projection methods used in the studies (Di Baldassarre, et al., 2011). It is disconcerting that most studies still rely on SRES emission scenarios and outdated climate modelling attempts and not on the Representative Concentration Pathways (RCPs) of the Fifth Assessment Report (IPCC, 2013), and on more recent climate modelling outcomes (see e.g. (Mastrandrea et al., 2011; Qin et al., 2016; Riediger et al., 2016; Rogelj et al., 2012). It remains uncertain whether in mid-century the Nile River will carry less or more water. The socio-economic implications are therefore impossible to predict.

There are significant variations in the findings of the economic valuation of the current water resources plans, especially when climate change is considered. For instance, one group of studies highlighted the great benefits of the Blue Nile cascade dams under historical conditions and that these benefits would increase under future climate conditions, whereas another group of studies finds that climate change would negatively influence these benefits.

One issue that is neglected in most studies is the heavy silt load in the Blue Nile and the impacts on reservoirs and irrigation canals. The current silt concentration during the flood season reach 10,000-14,000 ppm, five times more than in the 1920s. Reservoir operation affects and is affected by the sedimentation processes; therefore, considering reservoir sedimentation is important when studying the economic value of water resources developments in the basin. A second neglected issue is that the impacts of both unilateral and cooperative management for the entire basin system need to be accurately quantified. A third neglected issue is that certain components of the river basin have not been given attention, including water resources development effects on fisheries, navigation, and flood plain (recession) agriculture. In this dissertation, I considered reservoir sedimentation explicitly as well as the cooperative and non-cooperative management of the system.

7.2 THE EASTERN NILE MODELLING USING RIBASIM

The impacts of Eastern Nile water resources development options on the basin countries were assessed using the RIBASIM simulation model and through scenarios analysis. Different management options for dam operation were investigated including unilateral and cooperative transboundary management of dams. Unilateral management was modelled in RIBASIM by setting the source priority of border dams as empty, meaning that dams of an upstream country would not take into consideration the demand of a downstream country; water users of a downstream country thus cannot claim their demands from an upstream country's dams and will only use what is released from upstream dams in their own country. In case of cooperative management, each water user is connected to the upstream water supply infrastructure that can support its demand, even if this upstream supply infrastructure belongs to another riparian country.

In the study, the current operation rules of all existing reservoirs are assumed to remain the same. All dam developments are assumed online and at operational stage; and the transient stage (filling) and their short-term impacts have not been considered. In the initial condition of simulation, all reservoirs in the system are assumed full. The existing and proposed developments in Baro-Akobo-Sobat sub-basin have negligible effects on the system compared to the proposed large reservoirs in the other sub-basins and were therefore omitted. All assumptions related to the model and scenario development are detailed in chapter 4.

The findings indicate that new dam developments in Ethiopia would boost the hydropower generation in Ethiopia, particularly when following a cooperative management approach. Hydropower generation would increase in Sudan and insignificantly change in Egypt, when the system is operated in a cooperative manner. Development of Ethiopian dams show a small impact on Egypt hydropower generation because of operating AHD at lower water levels will result in lower evaporation losses from it. This finding is not new, but was already observed by Guariso & Whittington (1987) and others. Development of new irrigation projects would, however, reduce the hydropower generation of the three countries, but less than 15%. Power generation losses at AHD are very small due to dam developments in Ethiopia; however, power generation would be significantly reduced with the planned expansion of irrigation schemes in the upstream areas.

The development of the GERD in Ethiopia would (slightly) increase the supply reliability of existing irrigation projects in Sudan, but will slightly reduce if additional irrigation is developed. The supply-demand ratio of Sudanese irrigation projects would be reduced with the development of new irrigation projects under both cooperative transboundary and unilateral system management, with greater reductions in the latter. Full development of all planned dams in the basin would cause greater reductions in the supply-demand ratio for irrigation, because most new large dams are operated for hydropower generation.

Development of dams would also significantly affect the total net evaporation losses from reservoirs compared to the base scenario. While the basin-wide evaporation losses from reservoirs showed insignificant changes with the development of Ethiopian dams, the losses would increase with the development of the Main Nile dams in Sudan from $17.5 \times 10^9 \, m^3/yr$ to $22.5 \times 10^9 \, m^3/yr$.

As expected, the flow regime would be significantly influenced by dam and irrigation developments. Flows in the high flow season would decrease while they would increase during the low flow season. The average flows at AHD would decrease to 2,800 m^3/s from 5,400 m^3/s in the high flow season (3 months), and would increase from 1,500 m^3/s to 1,700 m^3/s in the low flow season (9 months) after GERD under cooperative management and assuming that the potential area of existing irrigation schemes are developed. Under non-cooperative management, the average flow in the high flow season would increase to 3,000 m^3/s while the average dry season flow will be 1,700 m^3/s, similar to the cooperative management condition. The results also reveal that the probability of Egypt not receiving its share of Nile water (inflows into AHD of $65.5 \times 10^9 \, m^3/yr$) would increase by the development of a cascade of hydropower dams upstream of GERD as well as new irrigation schemes including those in the Tekeze river, irrespective of the management conditions (cooperative or non-cooperative).

Managing the system unilaterally showed that, compared to cooperative system management, the generated power would increase in Ethiopia, and decrease in Sudan and Egypt by dam developments in Ethiopia, even without any further irrigation development. Power generation in Sudan and Egypt would, however, increase when the Main Nile dams in Sudan (four new dams) get operational. Irrigation development would generally decrease the generated hydropower from the proposed dams. Supply reliability of existing irrigation projects would not be affected by dam development until the development of the Main Nile dams in Sudan, when the reliability would reduce.

Most of the new large dams in the Eastern Nile are designed for hydropower generation (largely non-consumptive use of water). Results have therefore shown limited influence of dam developments and system management options on the inflow to AHD and thus hydropower-generation and downstream releases. So far, the Main Nile reservoirs in Sudan are planned for hydropower-generation only. This explains the increase of AHD hydropower generation by 10% in the unilateral compared to the cooperative management scenario, when the Main Nile dams have been developed, but without the full development of irrigation.

In conclusion, planning and managing the entire EN basin in a cooperative manner achieves benefits for all countries and reduces losses compared to the case of unilateral management, including evaporation losses and a reduction in supply reliability, provided that excessive irrigation development beyond sustainable levels of water availability is avoided. The location of planned large irrigation projects upstream of the proposed dams would reduce both the supply reliability of irrigation projects and the generated energy of the system. In addition, one may assume that unilateral management might also increase political tensions, which may lead to other types of losses, including economic.

The study shows that irrigation expansion would have a significant impact on the entire system. The study assumed fixed cropping patterns and fixed irrigated areas of the irrigation schemes. In reality, irrigable areas and cropping patterns vary seasonally based on inflow forecasts. It is recommended in future studies to assess the impact of different cropping patterns and irrigable areas of irrigation schemes on hydropower generation of the EN system, especially when the system is fully developed.

7.3 BENEFITS DISTRIBUTION OF WATER RESOURCES DEVELOPMENT FROM OPTIMAL OPERATION OF THE EASTERN NILE SYSTEM

In this part of the study, the operation of the Eastern Nile reservoirs system following the GERD development was optimized for both cooperative and non-cooperative management options. Two water uses were considered in the optimization, namely hydropower generation and irrigation. The model covered all existing hydraulic infrastructures in the EN and GERD. Only existing irrigation schemes plus those attached to Settit dam (168,000ha) are included in the analysis. The Eastern Nile system in Sudan was assumed to be constrained by the 1959 Agreement in all scenarios, which limits water withdrawals in Sudan to 18.5×10^9 m³/yr. The system was optimized on a monthly basis for a 7 year time horizon. The net economic return of irrigation and hydropower generation were assumed as 0.05 \$/m³ and 0.08 \$/kWh, respectively. The results showed there is no trade-off between hydro-energy and irrigation at the basin level when they are managed cooperatively (a 260 million $/year increase in hydro-generation would reduce irrigation returns by only 1 million $/year). A clear trade-off is, however, shown in case of non-cooperative system management: a 70 million $/year increase in hydro-generation would result in a 155 million $/year reduction of irrigation returns. Irrigation is more sensitive to the non-cooperative management scenario than hydro-energy, because the majority of irrigation lies in downstream countries, Sudan and Egypt. The results may encourage the riparian countries to cooperate, as the benefits would be more than when pursuing the non-cooperation option.

The findings support earlier studies that reported the positive impact of GERD development on the three Eastern Nile riparian (i.e., Ethiopia, Sudan and Egypt), if the three countries agreed to manage the Eastern Nile system cooperatively. The hydro-energy returns of all three countries would increase compared to the status quo, with Ethiopia witnessing the highest increase, as expected. Irrigation returns of Ethiopia and Egypt would remain as the status quo (100%), while Sudan would experience a reduction in the irrigation returns (−13%), because of increased demand by the irrigation scheme developed with the Settit Dam and the trade-off between irrigation schemes and downstream hydropower-demand of Merowe and AHD. Additionally, the basin-wide evaporation loss would reduce by the development of GERD because of storage reduction, and hence a reduction of water surface area, at AHD where the evaporation rate is higher compared to the location of GERD. As a result, more water would be available, which would allow for additional water uses and an increase of economic returns.

Non-cooperative system management would negatively impact the hydro-energy of Egypt over the cooperative management scenario (a reduction by 11%), without a significant increase in Ethiopian hydro-energy. Sudan hydropower generation is less sensitive (less than 2%) to system management scenarios because of its limited hydropower-generation capacity as indicated before. Irrigation in all countries showed a high sensitivity to the management scenarios, reducing the supply–demand ratio between 12% and 17% in all countries in the non-cooperation scenario. Along with the non-cooperative management, the reduction in irrigation supply is attributed to the presence of a trade-off between hydropower generation and irrigation within each of the two countries of Ethiopia and Sudan. This is because in both countries, hydropower dams are located downstream of the irrigation schemes.

It should be noted that the model does not include flow routing, and therefore, cannot handle flood management. The study has not covered other impacts of GERD, such as on sediment management, recession agriculture, and other environmental impacts. The economic value of water in transboundary rivers is dynamic and varies according to the type (consumptive and non-consumptive) and location (upstream and downstream) of users. The economic returns of hydropower and irrigation were, however, assumed to be the same for the three countries; as such, an in-depth analysis of the water value at the macro-economic scale is beyond the scope of the study. Kahsay (2017) assessed the impact of GERD on the economy of the Eastern Nile countries using a combined hydro-economic optimization model to determine the optimal water allocation, and computable general equilibrium (CGE) to simulate the impact of optimization decisions on the economy of the countries. The findings indicated that the GERD would generate basin-wide economic benefits in the EN basin.

7.4 DEVELOPMENT OF THE EASTERN NILE RESERVOIR SYSTEM SEDIMENTATION MODEL - INCLUDING AN APPLICATION TO ROSEIRES DAM

Three scenarios were developed for optimizing the operation of Roseires reservoir during the flood season with high sediment load. The scenarios include maximizing releases for irrigation and hydropower generation, maximizing sediment releases, and a combination of both. The operation was optimized for only irrigation and hydropower generation during the dry season in all three scenarios.

The findings showed that including sediment management as an objective in reservoir operation optimization would positively impact the net economic returns of hydropower and irrigation in the long term. The average annual economic return of hydropower generation, irrigation and sediment management return would be 136, 549 and 77 million USD, respectively, compared to 128, 520 and 38 million USD when sediment management is ignored.

Downstream water requirement and irrigation demand would be satisfied (100%) when sediment management is adopted (optimized) in the operation. The average supply of irrigation schemes would reach 98% of the minimum requirement when optimization of reservoir operation considers only irrigation and hydropower, with deficits occurring 10% of the time. The life time of the reservoir would be much shorter in case of not including sediment management.

In comparison with the observed operation in 1995, the analysis shows that sediment deposition can be further reduced which has a positive impact on hydropower generation and irrigation supply. The findings indicate that reservoir sustainability can be enhanced by modifying the current operation rule, which suggest lowering the minimum operation water level by 0.25 meter to 466.75 m.a.s.l. during July and August, and increasing levels compared to the current rule from December through June.

The modelling results show that changes in the assumed value of 1 unit of sediment released from the reservoir hardly influences the amount of sediment released, the volume of irrigation water supplied or the amount of generated energy. This finding may be explained by that sediment management is conducted in only three months of the year, and part of the sediment is released with the water running through the turbines. Increasing the value of sediments flushed would also not significantly change water levels, because water levels are as close to the minimum operation level during the months when sediment management is carried out.

The findings of this study are based on priority given to irrigation demands over hydropower generation, similar to the current operation of Roseires reservoir. The results show that sediment management does not conflict with water use for irrigation. The results might change when hydropower is included with equal or higher priority over irrigation. In this case, less water is expected to be released to create head for hydropower generation during the flood season, which conflicts with sediment management.

Different combinations of multiple objectives of reservoir operation were investigated in this study, and equal priorities were given to objectives when multiple objectives were considered, except to the objective of hydropower generation. Further investigation is recommended for conditions where the three objectives are given different weights, to assess the sensitivity of the findings to changes in the priority of each objective.

The trap efficiency model was used to simulate sediment dynamics in this study for optimizing the current operation of Roseires Dam. The model simplifies the complex process of sediment transport. Therefore, further studies are recommended that use advanced sediment transport models to optimize the operation. Different sediment management options are also recommended to be evaluated to assess the feasibility of the currently used method over other options.

Although the method adopted in this study is primarily developed for multiple reservoir system, it has been applied for Roseires dam, a single reservoir, because of data availability at the time of the study. It is recommended to apply this method in future studies for multiple reservoirs to address sediment management that has so far largely been ignored in Easter Nile system modelling studies. In particular, the impact of the GERD on the sediment dynamics in downstream reservoirs, and thus their operation, also needs to be studied urgently. The GERD is anticipated to trap most sediment that otherwise would have entered Roseires reservoir, which could create the opportunity to orient operation of the latter towards its main economic objectives, namely irrigation water supply and hydropower generation, generating additional benefits.

7.5 SUMMARY AND RECOMMENDATIONS

This study presented a comparative quantification of the impacts of water resources development in the Eastern Nile and explored several system management options at both regional and country levels. In addition, the study proposed new operation rules for improving operation of the current system when new infrastructures are developed and operated either unilaterally or cooperatively. The distribution of the benefits between countries was quantified for both cooperative and non-cooperative management options. As a result, it is shown that developing a collaborative and unified perspective of the countries towards new projects can be beneficial for all.

Unlike the current operation, optimal operation of the system for hydropower generation and irrigation following infrastructure development would shift towards hydropower generation. This shift is attributed to many interrelated aspects that need to be explored more in future studies, such as the largely non-consumptive nature of hydropower, its relatively high economic return and the location of hydropower dams in the basin.

The current unilaterally developed plans for water resources development are in conflict, both basin-wide between riparian countries and within each of the riparian countries. These plans therefore need to be reconsidered with a focus on basin-wide improved use of the available water and on collective benefits. Locating hydropower dams downstream of large irrigation projects would reduce both the generated energy of the system and the supply reliability of irrigation projects. Among other water users, irrigation projects have shown a greater sensitivity to non-cooperative management of the system. Further studies could include the impact of varying cropping patterns and irrigated areas of irrigation projects. Future studies should also include the Main Nile system downstream of Aswan High Dam.

The operation of reservoirs can be optimised further when sediment management is included. Advanced sediment transport models are recommended to be used for sediment management simulations in the EN multi-reservoir system management. Trap efficiency models can be used for planned dams that lack observed data, while sediment transport models can be calibrated and more accurately estimate the trap efficiency of existing reservoirs.

8

APPENDICES

8.1 APPENDIX-I STUDY AREA

Table 1 Major Reservoirs, Hydropower Plants and Irrigation Projects in the Eastern Nile
(Source: Verhoeven (2011); Goor (2010); ENTRO (2007))

Project name	Status	Current capacity [Potential capacity] (MW)	Reservoir capacity [Potential capacity] (m^3) Irrigation area (ha)
Ethiopia			
Tekeze River			
Tekeze	Operating since 2009	300	3.1×10^9 45,000
Tekeze II	Proposed, 2020 Expected year of commission	[450]	Not Available
Lake Tana Tributaries			
Tana Beles (Lake Tana-Beles River Transfer)	Operating since 2010	460	9.12×10^9 [140,000-150,000]
Abbay(Blue Nile)			
Tis Abbay I, Abbay River	Operating since 1964	11.4	[50,000]
Tis Abbay II, Abbay River	Operating since 2001	68-85	
Fincha' a , Fincha' a River	Operating since 1973, Extra unit added and commissioned 2006	128-134	460×10^6 - 2.4×10^9
Fincha'a-Amerti-Neshi, Fincha' a River	Under construction, 57% completed as of April 2011	[97]	-
Grand Ethiopian Renaissance Dam, Blue Nile	Under construction, started April 2011, expected complete date after 2017	[5,250]	[63-67 $\times 10^9$]
Chemoga- Yeda Hydropower Project, including dams on Chemoga, Yeda, Sens, Getla, Bogrna	Construction contract signed. Expected completion of Phase I in 2015.	[278]	-
Jema, Jema River	Proposed, Feasibility study complete	-	[173 $\times 10^6$] [7,800]
Mabil, Blue Nile (replaced by Beko Abo Dam)	Proposed, 2021Expected year of commission	[1,200]	[13.6 $\times 10^9$]
Mendaya, Blue Nile	Proposed under ENSAP, Nile Basin Initiative , 2030 Expected year of commission	[1,620-2,000]	[13 $\times 10^6$ -15.9 $\times 10^6$]
Beko Abo, Blue Nile	Proposed under ENSAP, Nile Basin Initiative.	[2,100]	10.5 $\times 10^6$

Project name	Status	Current capacity [Potential capacity] (MW)	Reservoir capacity [Potential capacity] (m³) Irrigation area (ha)
Border, Blue Nile (replace d by GERD)	Proposed under ENSAP, Nile Basin Initiative, 2026 Expected year of commission	[800-1,400]	[11.1 x10^9]
Karadobi, Blue Nile	Proposed under ENSAP, Nile Basin Initiative, 2023 Expected year of commission	[1000-1,600]	[32.5- 41 x10^9]
Diddessa irrigation project, including dams on Diddessa, Dabana, Negeso	Proposed, 2038 Expected year of commission	[308- 615]	[55,000]
Anger- Nekemte Irrigation Project, including dams on Anger, Nekemte	Proposed, 2038 Expected year of commission	[15-20]	[26,000]
Dabus, Dabus River	Proposed, feasibility studies ongoing	[425]	-
Baro River and its tributaries			
Sor, tributary of Geba	Operating since 1990	5	-
Alwero Irrigation Project, Alwero river	Operating since 1995	Not Available	74,600
BaroI and II, Baro River	Proposed under ENSAP, Nile Basin Initiative, 2034 Expected year of commission	[850-896]	-
Geba I and II, Geba River	Proposed under ENSAP, Nile Basin Initiative, 2016 Expected year of commission	[254 - 366]	-
Birbir A and B	Proposed, feasibility studies ongoing	[467 - 508]	-
Tams	Proposed, feasibility studies ongoing	[1,000]	-
Sudan **Main Nile**			
Merowe, 4th Cataract, Nile	Operating since 2009	1,250 [2,000]	12.5 x10^9 [380,000]
Kajbar, 3rd Cataract, Nile	Under construction, 2016 Expected year of commission	[300–360]	8.2 x10^6

Project name	Status	Current capacity [Potential capacity] (MW)	Reservoir capacity [Potential capacity] (m³) Irrigation area (ha)
Shereik ,3rd Cataract, Nile	Construction contract signed	[315–420]	-
Dal ,2nd Cataract, Nile	Proposed, Feasibility studies ongoing	[340–600]	-
Mograt ,4th Cataract, Nile	Proposed, Feasibility studies	[240-312]	-
Dagash, Main Nile	complete	[285-312]	-
Sabaloka, 6th Cataract, Nile		[120-205]	[4×10^9]
Atbara River and tributaries			
Khashm Elgirba Atbara River	Operating since 1964	0–7 [12.5]	1.3×10^9 206,640
Upper Atbara Project, including Rumela Dam in Atbara River, Burdana Dam in Settit River	Under construction, completed in 2015	Rumela [120] Burdana [15]	[2.7×10^9] Rumela [190,000] Burdana [210,000]
Blue Nile			
Roseires Dam, Blue Nile	Operating since1966; 1971 Hydropower plant added; 2013 d completion of dam heightening	100–250 [275]	2.2×10^9 [$3.7–4 \times 10^9$] 1.7×10^6
Sennar Dam, Blue Nile	Operating since 1925; 1962 Hydropower plant added; Rehabilitation planning ongoing	15 [45]	930×10^6 870,750
Egypt **Main Nile**			
Aswan High Dam	Operating	2100	$132,00 \times 10^6$
Old Aswan Dam	Operating	500	0(run of river) No irrigation
Esna	Operating	90	0(run of river) No irrigation

Table 1 Summary of reviewed studies on Nile river basin simulation

Simulation models

Study/Authors (details are in the reference list)	Basin	Single/multi-reservoirs.	Purpose of study	Model	Reservoir sedimentation	Economic Assessment	Climate change impact	Cooperative/non Cooperative (national) management	Transient Stage (Filling)/ Steady State condition(operational)
Nile Valley Plan (NVP), Morrice and Allan, 1958	Nile basin	Multi-reservoirs	planning reservoirs	Hydraulic and hydrologic simulation	N	N	N	Cooperative management is assumed	Steady state
USBR, 1964	Upper Blue Nile	Multi-reservoirs	planning reservoirs	Simple routing calculations	Y	Y	N	Cooperative management is assumed	Steady state
NILESIM, Levy and Baecher, 1999	Nile basin	Multi-reservoirs	existing and planning reservoirs	Hydrologic simulation	N	N	N	N/A	Steady state
ENTRO-ENPMP, ENTRO 2012	E.Nile basin	Multi-reservoirs	existing and planning reservoirs	SWAT, RiverWare, RIBASIM, MIKEBASIN, HEC-ResSim	N	Y	N	N/A	Filling and steady state
Wheeler et al., 2016	E.Nile basin	Multi-reservoirs	existing and planning reservoirs	RiverWare	N	N	N	Cooperative management is assumed	Filling stage

Study/Authors (details are in the reference list)	Basin	Single/multi-reservoirs.	Purpose of study	Model	Reservoir sedimentation	Economic Assessment	Climate change impact	Cooperative/non Cooperative (national) management	Transient Stage (Filling)/ Steady State condition(operational)
Zhang et al.,	Blue Nile	Multi-reservoirs	existing and planning reservoirs	Hydrologic simulation and Reservoir and hydropower model	N	N	N	N/A	Filling stage
Mulat and Moges, 2014a	Blue Nile	Multi-reservoirs (two), Ethiopia and AHD	existing and planning reservoirs	MIKE BASIN	N	N	N	N/A	Filling and steady state
Mulat and Moges, 2014b	Blue Nile	Multi-reservoirs	existing and planning reservoirs	MIKE BASIN	N	N	N	N/A	Filling state
Mohamed, 1990	Blue Nile	Multi-reservoirs (two), Sudan	existing reservoir management	Water budget balance based simulation model	Y	Y	N	N/A	Steady state
Abdallah and Stamm , 2013	Blue Nile	Multi-reservoirs (two), Sudan	existing reservoir management	Fuzzy set theory	Y	N	N	N/A	Steady state
Hurst 1966	Main Nile	Single reservoir-AHD	existing reservoir management	Simple routing calculations	N	N	N	N/A	Steady state

Study/Authors (details are in the reference list)	Basin	Single/multi-reservoirs.	Purpose of study	Model	Reservoir sedimentation	Economic Assessment	Climate change impact	Cooperative/non Cooperative (national) management	Transient Stage (Filling) / Steady State condition(operational)
Wassie, 2008	Lake Tana	Natural Lake	Natural Lake	WAFLEX	N	N	N	N/A	Steady state
Abreha, 2010	Tekeze river	Single reservoir-TK5	existing reservoir management	RIBASIM, Desktop Reverse model	N	N	N	N/A	Steady state
Jeuland, 2010	Blue Nile	Multi-reservoirs (two)	existing and planning reservoirs	Hydrologic simulation	N	Y	Y	N/A	Steady state
McCartney and Menker Girma 2012	Upper Blue Nile	Multi-reservoirs	existing and planning reservoirs	Dynamic climate model, SWAT, WEAP	N	N	Y	N/A	Steady state
Jeuland and Whittington, 2014	Upper Blue Nile	Multi-reservoirs	Planning reservoir	Hydrologic simulation,	N	Y	Y	cooperative and non-cooperative management are investigated	Steady state
King and Block 2014	Blue Nile	Single reservoir-GERD	Planning reservoir	Hydrologic simulation and Reservoir and hydropower model	N	N	Y	N/A	Filling stage

Study/Authors (details are in the reference list)	Basin	Single/multi-reservoirs.	Purpose of study	Model	Reservoir sedimentation	Economic Assessment	Climate change impact	Cooperative/non Cooperative (national) management	Transient Stage (Filling)/ Steady State condition(operational)
Wondimagegne hu and Tadele, 2015	Upper Blue Nile	Multi-reservoirs	Planning reservoir	Regional climate model, HEC-HMS, HEC-ReSim	N	N	Y	N/A	Steady state
Optimization models									
Guariso et al., 1981	Main Nile	Single reservoir-AHD	existing reservoir manageme nt	Stochastic NLP	N	N	N	N/A	Steady state
Stedinger,Sule et al., 1984	Main Nile	Single reservoir-AHD	existing reservoir manageme nt	Stochastic DP	N	N	N	N/A	Steady state
Kelman, Stedinger et al.,1990	Main Nile	Single reservoir-AHD	existing reservoir manageme nt	Sampling stochastic DP	N	N	N	N/A	Steady state
Ghany, 1994	Blue Nile	Multi-reservoirs (two), Sudan	existing reservoir manageme nt	Stochastic DP	N	N	N	N/A	Steady state
Guariso and Whittington, 1987	E.Nile basin	Multi-reservoirs	existing and planning reservoirs	LP	N	N	N	Cooperative management is assumed	Steady state

Study/Authors (details are in the reference list)	Basin	Single/multi-reservoirs.	Purpose of study	Model	Reservoir sedimentation	Economic Assessment	Climate change impact	Cooperative/non Cooperative (national) management	Transient Stage (Filling)/ Steady State condition(operational)
Whittington et al., 2005	Nile basin	Multi-reservoir	existing and planning reservoirs	Deterministic NLP	N	Y	N	cooperative and non-cooperative management are investigated	Steady state
Goor et al., 2010	E.Nile basin	Multi-reservoir	existing and planning reservoirs	SSDP	N	Y	N	Cooperative management is assumed	Steady state
Daine et al., 2014	E.Nile basin	Multi-reservoir	existing and planning reservoirs	SSDP	N	Y	N	non-cooperative management are investigated	Steady state
Lee et al., 2012	Main Nile	Single reservoir-AHD	existing reservoir management	DP	Y	Y	N	cooperative and non-cooperative management are investigated	Steady state
Satti et al., 2014	Blue Nile and Main Nile	Multi-reservoir	existing and	GAMS	N	Y	Y	N/A	Steady state

Study/Authors (details are in the reference list)	Basin	Single/multi-reservoirs.	Purpose of study	Model	Reservoir sedimentation	Economic Assessment	Climate change impact	Cooperative/non Cooperative (national) Cooperative management	Transient Stage (Filling)/ Steady State condition (operational)
			planning reservoirs						
Block and Rajagopalan, 2007	Upper Blue Nile	Multi-reservoir	planning reservoir	Standalone optimization model	N	Y	Y	N/A	Filling stage
Block and Strzepek, 2010 Block and Strzepek, 2012	Upper Blue Nile	Multi-reservoir	planning reservoir	Standalone optimization model	N	Y	Y	N/A	Filling stage
Combined simulation and optimization models									
Musa, 1985	Blue Nile	Multi-reservoir (two), Sudan	existing reservoir management	Simulation: MODSIM; Optimization: Heuristic optimization, LP	N	N	N	N/A	Steady state
Hamad, 1993	Blue Nile	Multi-reservoir (two), Sudan	existing reservoir management	Simulation: IRIS; Optimization: Slice method	N	N	N	N/A	Steady state
Hassaballah, 2010	Blue Nile	Multi-reservoir (two)	existing and planning reservoirs	Simulation: MIKEBASIN; Optimization: NSGA-II	N	N	N	Cooperative and non-cooperative management is assumed	Filling stage
Nile River basin Decision Support Tool	Nile basin	Multi-reservoir	existing and	Simulation: River and reservoir routing models; Optimization: ELQGC	N	N	N	Cooperative management is assumed	Steady state

Study/Authors (details are in the reference list)	Basin	Single/multi-reservoirs.	Purpose of study	Model		Reservoir sedimentation	Economic Assessment	Climate change impact	Cooperative (national)/non Cooperative (national) management	Transient Stage (Filling)/ Steady State condition(operational)
(DST), LVDST Georgakakos, 2006 and 2007			planning reservoirs							
Nile Basin Decision Support System (NBDSS), Hamid, 2013	Nile basin	Multi-reservoir	existing and planning reservoirs	Simulation	MIKEBASIN, WEAP	N	Y	N	Cooperative management is assumed	Steady state
				Optimization	NSGA-II, Suffled Complex Evolution (SCE) with its variant Simplex (only one complex), Monte-Carlo Simulation, Dynamically Dimensioned Search (DDS)					

8.3 APPENDIX-III EASTERN NILE MODELLING USING RIBASIM

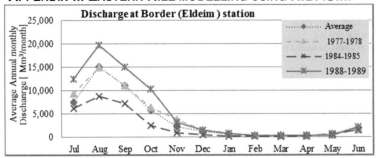

Figure 1: Average annual monthly discharge (July- June) at Border (Eldiem) station

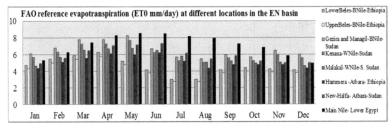

Figure 2 The monthly reference evapo-transpiration (ET0) at different locations in the Eastern Nile basin

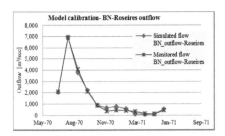
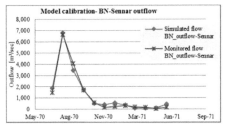

Figure 3 Measured and Simulated downstream flow (m³/sec) at Roseires and Sennar dams (Jul 1970-Jul 1971)

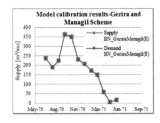
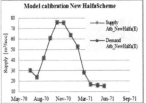
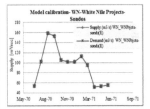

Figure 4 Demand (m³/sec) and the Supply (m³/sec) of Gezira and Managil, New Halfa and White Nile Irrigation Project (Jul 1970-Jul 1971)

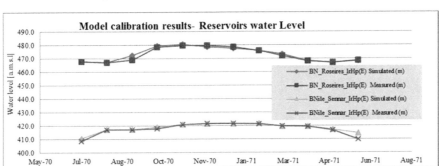

Figure 5 Measured and Simulated water levels (a.m.s.l) at Roseires and Sennar dams (Jul 1970-Jul 1971)

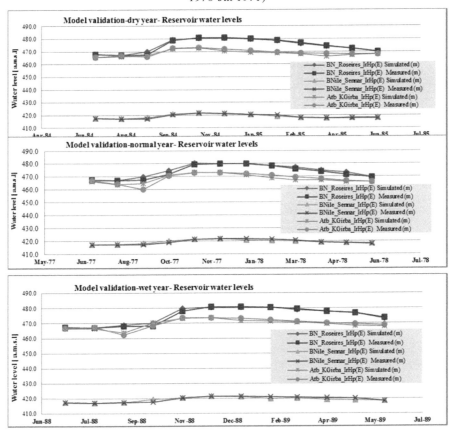

Figure 6 Measured and Simulated water levels (a.m.s.l) of Roseires, Sennar and Khashm Elgirba dams at years of different hydrologic conditions: dry (Jul 1984-Jun 1985), normal (Jul 1977-Jun 1978) and wet (Jul 1988-Jun1989)

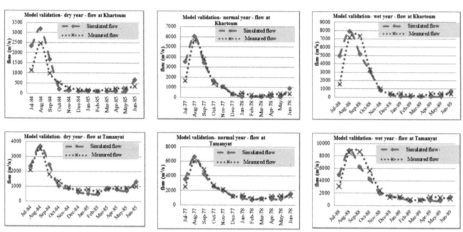

Figure 7 Measured and simulated flow at key locations in the Blue Nile, Atbara River and the main

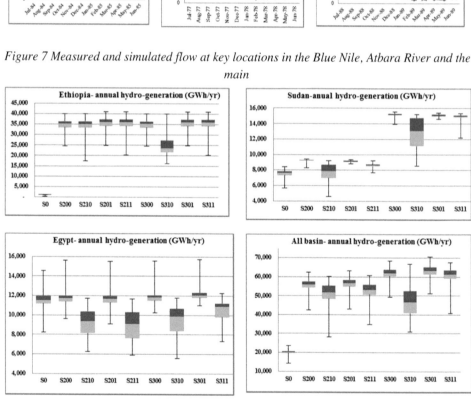

Figure 8: Box plot of the annual generated energy (GWh/year) of the basin countries for Ethiopian dam(S2xx) and full basin (S3xx) development scenarios, with (Sx0x) and without

(Sx1x) irrigation development in case of cooperative transboundary (Sxx0) and unilateral (Sxx1) system management

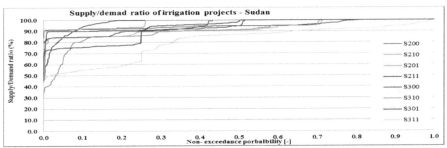

Figure 9: Non- exceedance probability of the average monthly supply to demand ratio (%) of Sudan existing (Sx0x) and potential (Sx1x) irrigation projects after Ethiopian dams (S2xx) and basin full (S3xx) development under cooperative transboundary management (Sxx0), unilateral management (Sxx1) and Base Scenario (S0)

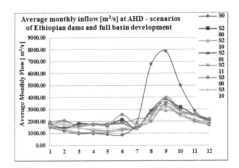
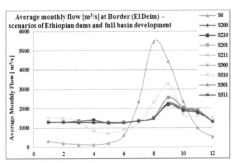

Figure 10 Average monthly flow [m3/s] at (a) AHD and (b) Border after Ethiopian dams (S2xx) and basin full (S3xx) development, with existing (Sx0x) and potential (Sx1x) irrigation projects under cooperative transboundary (Sxx0) and unilateral (Sxx1) system management

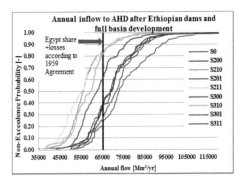
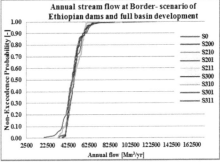

133

*Figure 11 Cumulative distribution function (CDF) of the annual stream flow at (a) AHD and
(b) Border after Ethiopian dams (S2xx) and basin full (S3xx) development, with existing
(Sx0x) and potential (Sx1x) irrigation projects under cooperative transboundary (Sxx0) and
unilateral (Sxx1)system management*

Table 1 Hedging rules for model calibration

Storage zones between firm and dead storage	Lower boundary of zone [% between firm and dead storage]	Water allocation [% of target release]
-	100	-
Zone 1	80	90
Zone 2	60	70
Zone 3	40	50
Zone 4	20	30
Zone 5	0	10

Table 2 Irrigation Projects data used for model calibration

Month	BN_GeziraMenagil(E) Mm³/month[1]	Atb_NewHalfa(E) Mm³/month[3]	WN_WNPrjcts-sonds(E) Mm³/month[3]
Jul-1970	640.66	81.45	144.61
Aug-1970	507.00	63.70	274.93
Sep-1970	586.54	109.14	411.54
Oct-1970	974.82	163.42	411.54
Nov-1970	912.15	197.05	274.36
Dec-1970	622.23	203.10	274.36
Jan-1971	555.39	171.25	274.36
Feb-1971	418.39	128.27	274.14
Mar-1971	403.29	75.70	257.21
Apr-1971	153.88	44.46	133.86
May-1971	18.44	43.86	142.32
Jun-1971	49.06	40.42	143.26

1: Source: Estimated from Roseires Heightening Report (McLellan, 1987) and MWRE-Dams Operation Department

3: Source: Nile Water Master Plan (MOI, 1979) and MWRE-Nile Water Directorate 2014

Table 3Irrigation Projects data used for model validation- Dry Year (July 1984-June1985)

Month	BN_GeziraMenagil(E) Mm³/month[1]	BN_USennarRahad-I(E) Mm³/month[1]	BN_UpSennar(E) Mm³/month[2]	BN_GinaidBNpumps(E) Mm³/month[2]	Atb_NewHalfa(E) Mm³/month[3]	WN_AsalyaSuger (E) Mm³/month[3]	WN_Kenana-I(E) Mm³/month[2]	WN_WNPrjcts-sonds(E) Mm³/month[3]
Jul-1984	559.70	112.27	63.72	133.96	133.96	17.00	89.81	176.43

134

Month	BN_Gezir aMenagil(E)	BN_USe nnarRa had-I(E)	BN_UpS ennar (E)	BN_Gin aidBNp umps(E)	Atb_Ne wHalfa (E)	WN_As alyaSu ger (E)	WN_Ke nana-I(E)	WN_W NPrjcts -sonds (E)
	$Mm^3/$ month[1]	$Mm^3/$ month[1]	$Mm^3/$ month[2]	$Mm^3/$ month[2]	$Mm^3/$ month[3]	$Mm^3/$ month[3]	$Mm^3/$ month[2]	$Mm^3/$ month[3]
Aug-1984	873.47	192.49	85.30	186.18	186.18	22.67	72.14	224.03
Sep-1984	872.11	202.17	84.10	222.35	222.35	17.35	68.79	307.64
Oct-1984	827.91	172.95	74.55	179.02	179.02	21.15	63.11	295.95
Nov-1984	585.78	122.18	95.58	133.24	133.24	21.04	81.67	298.38
Dec-1984	482.18	109.33	83.62	116.40	116.40	17.23	69.97	283.95
Jan-1985	368.32	95.57	62.56	72.99	72.99	15.71	63.55	299.73
Feb-1985	337.06	85.93	57.54	37.05	37.05	19.32	66.54	145.37
Mar-1985	78.26	13.86	37.70	37.49	37.49	18.05	60.10	126.84
Apr-1985	31.10	0.00	54.96	41.66	41.66	16.86	90.56	114.74
May-1985	32.96	0.00	36.39	80.25	80.25	19.94	88.56	116.67
Jun-1985	332.42	136.88	36.76	165.54	165.54	19.95	77.30	143.54

1: Source: Roseires Heightening Report (McLellan, 1987)
2: Source: Long term Power plan and MWRE-Nile Water Directorate 2014
3: Source: Nile Water Master Plan (MOI, 1979) and MWRE-Nile Water Directorate 2014

Table 4 Irrigation Projects data used for model validation- Normal Year (July 1977-June1978)

Month	BN_Gezi raMenag il (E)	BN_USen narRahad -I (E)	BN_UpS ennar (E)	BN_Ginaid BNpumps(E)	Atb_New Halfa (E)	WN_W NPrjcts-sonds (E)
	$Mm^3/$ month[1]	$Mm^3/$ month[1]	$Mm^3/$ month[2]	$Mm^3/$ month[2]	$Mm^3/$ month[3]	$Mm^3/$ month[3]
Jul-1977	529.28	54.20	86.54	35.06	122.08	166.13
Aug-1977	856.10	14.92	8.87	22.82	68.58	210.95
Sep-1977	870.73	67.98	42.74	36.30	172.28	280.34

135

Oct-1977	862.89	70.95	105.64	42.30	229.65	278.67
Nov-1977	809.55	59.76	106.58	38.97	225.24	271.89
Dec-1977	814.60	51.40	83.08	37.37	199.84	267.37
Jan-1978	803.71	36.20	73.21	34.21	167.84	267.37
Feb-1978	449.57	29.40	107.21	32.91	132.60	117.13
Mar-1978	88.33	5.25	55.40	38.58	72.49	113.15
Apr-1978	36.55	0.00	39.15	25.74	43.80	99.05
May-1978	38.73	0.00	35.99	22.69	45.05	100.72
Jun-1978	661.43	1.25	34.56	30.86	83.99	128.04

1: Source: Roseires Heightening Report (McLellan, 1987)
2: Source: Long term Power plan and MWRE-Nile Water Directorate 2014
3: Source: Nile Water Master Plan (MOI, 1979) and MWRE-Nile Water Directorate 2014

Table 5 Irrigation Projects data used for model validation- Wet Year (July 1988-June1989)

Month	BN_GeziraMenagil (E)	BN_USennarRahad-I(E)	BN_UpSennar (E)	BN_GinaidBNpumps (E)	Atb_NewHalfa(E)	WN_AsalyaSuger (E)	WN_Kenana-I(E)	WN_WNPrjcts-sonds (E)
	$Mm^3/$month[1]	$Mm^3/$month[1]	$Mm^3/$month[2]	$Mm^3/$month[2]	$Mm^3/$month[3]	$Mm^3/$month[3]	$Mm^3/$month[2]	$Mm^3/$month[3]
Jul-1988	525.00	75.60	32.05	22.61	173.67	34.70	65.41	176.10
Aug-1988	477.49	30.10	49.52	8.80	16.23	61.60	55.10	196.20
Sep-1988	512.95	181.04	45.50	37.94	99.46	8.80	56.60	217.00
Oct-1988	824.10	150.50	85.36	28.32	218.80	12.97	62.10	250.00
Nov-1988	779.03	180.83	106.83	29.30	206.87	20.82	81.40	248.31
Dec-1988	775.60	115.30	55.56	27.04	183.20	16.62	70.20	250.15
Jan-1989	564.90	70.14	57.43	23.02	143.40	14.67	64.19	270.35
Feb-1989	545.71	55.80	52.74	20.39	124.33	15.62	67.88	101.41
Mar-1989	363.10	30.15	43.48	20.26	95.00	15.00	61.47	110.50
Apr-1989	88.45	39.58	40.34	23.44	55.86	19.58	89.07	99.36

May-1989	69.75	31.30	59.80	21.83	52.88	25.00	89.00	96.75
Jun-1989	**195.71**	**59.02**	**99.30**	**30.55**	**52.29**	**19.57**	**79.26**	**124.41**

1: Source: Estimated from Roseires Heightening Report (McLellan, 1987) and MWRE-Nile Water Directorate 2014

2: Source: Long term Power plan and MWRE-Nile Water Directorate 2014

3: Source: Nile Water Master Plan (MOI, 1979) and MWRE-Nile Water Directorate 2014

Table 6 Reservoir Level- Area- Volume data used for calibration and validation

Process	Year	Roseires	Sennar	K.Girba	J.Aulia
Calibration	Jul1970-Jun1971	1966 Bathymetric data	1925 Bathymetric data	1964 Bathymetric data	1937 Bathymetric data
Validation	Jul1977-Jun1978	1966 Bathymetric data	1925 Bathymetric data	1964 Bathymetric data	1937 Bathymetric data
	Jul1984-Jun1985	1985 Bathymetric data	1985 Bathymetric data	1978 Bathymetric data	1937 Bathymetric data
	Jul1988-Jun1989	1985 Bathymetric data	1985 Bathymetric data	1978 Bathymetric data	1937 Bathymetric data

137

8.4 APPENDIX-IV EASTERN NILE RESERVOIRS SYSTEM OPTIMIZATION

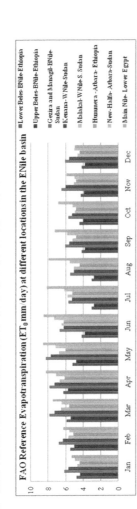

Figure A1. The monthly reference evapo-transpiration (ET₀) at different locations in the Eastern Nile Basin. Source: Digna et al., 2018a

Table A1. Major reservoirs, hydropower plants and irrigation projects in the Eastern Nile. Sources: Verhoeven, (2011); Goor, (2010); ENTRO. (2007); Van der Krogt and Ogink, (2013)

Country	River/Project Name	Status	Hydropower Capacity in 2017 (Potential Capacity) (MW)	Reservoir Capacity in 2017 (Potential Capacity) (m³)	Irrigation Area in 2017 (Potential Irrigated Area) (ha)
Ethiopia	Tekeze				
	Tekeze V	Operating since 2009	300	9.3×10^9	45,000
	Lake Tana tributaries				
	Tana–Beles River Transfer	Operating since 2010	460	9.1×10^9 (volume of water from Lake Tana for power production)	140,000
	Abbay (Blue Nile)				

Country	River/Project Name	Status	Hydropower Capacity in 2017 (Potential Capacity) (MW)	Reservoir Capacity in 2017 (Potential Capacity) (m³)	Irrigation Area in 2017 (Potential Irrigated Area) (ha)
	Tis Abbay I	Operating since 1964	11.4		50,000
	Tis Abbay II	Operating since 2001	68–85		
	GERD	Under construction	(6450)	(74 × 10⁹)	
	Baro River and tributaries Sor, tributary of Geba	Operating since 1990	5		
	Alwero Irrigation Project, Alwero river	Operating since 1995	N/A		74,600
	Barol and II, Baro River	Proposed under ENSAP, NBI	(850–896)		
	Geba I and II, Geba River	Proposed under ENSAP, NBI	(254–366)		
	Birbir A and B	Proposed, feasibility studies ongoing	(467–508)		
	Tams	Proposed, feasibility studies ongoing	(1000)		
Sudan	Atbara and tributaries				
	Khashm Elgirba	Operating since 1964	7 (12.5)	1.3 × 10⁹	206,600
	Rumela and Burdana Complex Dam in Settit River	Operating since 2016	320	2.7 × 10⁹	(300,000)
	Blue Nile				
	Roseires Dam	Operating since 1966; Dam heightening completed in 2013	100 (275)	2.2 × 10⁹ 3.7 × 10⁹	

Country	River/Project Name	Status	Hydropower Capacity in 2017 (Potential Capacity) (MW)	Reservoir Capacity in 2017 (Potential Capacity) (m^3)	Irrigation Area in 2017 (Potential Irrigated Area) (ha)
	Sennar Dam, White Nile	Operating since 1925; Rehabilitation—ongoing	15 (45)	0.6×10^9 (0.9×10^9)	870,700
	Jebel Aulia, White Nile	Operated since 1937 Rehabilitated in 2005	30.4-35	3.5×10^9	152,300
	Merowe, 4th Cataract, Nile	Operating since 2009	1250 (2000)	12.5×10^9	380,000
Egypt	Main Nile				
	Aswan High Dam	Operating	2100	162×10^9	
	Old Aswan Dam	Operating	500		
	Esna	Operating	90		
	Assyut	Operating	(32)		690,000
	Delta	Operating	---		305,000
	Naga Hammadi	Operating	64		320,000

8.5 APPENDIX-V RESERVOIR SEDIMENTATION

Table 1. Literature review summary

No	Study	Country/Region/Basin	Single/Multiple reservoirs	Existing/Planned/Combined	Optimization/Simulation	System condition
1	(González Iñiguez, 2017)	Ecuador	Single	Existing	Opt	GA: Max Hp and the max storage
					Sim	Tsingha University for flushing
2	(Greg Schellenberg et al., 2017)	N/A	Review paper	Existing	Opt	N/A
					Sim	MIKE 21 + Sluicing
3	Rashid - 2015	Pakistan	Multiple	Existing	Opt	Hp, Irrigation, flood and Sediment
					Sim	Tsingha University for flushing
4	(Afrin, 2014)	USA	Single	Existing	Opt	N/A
					Sim	MIKE 11+ Sluice gate
5	(Hajiabadi & Zarghami, 2014; Shokri et al., 2013)	Iran	Single	Existing	Opt	GA
					Sim	Flushing simulation-Tsinghau University method
6	(Shokri, et al., 2013)	Iran	Single	Existing	Opt	SDP
					Sim	Flushing simulation-Tsinghau University method
7	(Lewis, et al., 2013)	Australia	Single	Existing	Opt	N/A
					Sim	Estimate TE for Tropical Areas. Used Brune and Calibrated to for Existing reservoirs.
8	(Lee & Foster, 2013)	USA	Single	Existing	Opt	N/A
					Sim	CE-QUAL-W2 V3.6 is a two-dimensional, hydrodynamic water quality model
9	(Borji, 2013)	Blue Nile-GERD	Single	Planned	Opt	N/A
					Sim	RASCON –Assessed all empirical methods====then studied different management measures (Theoretical based Analysis)

No	Study	Country/Region/Basin	Single/Multiple reservoirs	Existing/Planned/Combined	Optimization/Simulation	System condition
10	(Tebbi et al., 2012)	Algeria	Single	Existing	Opt	GA-Optimizing the Cumulative sediment trapped and that from bathymetric survey (6 bathymetric surveys) ====Nash criterion method.
					Sim	Simulate sediment management such as dredging.
11	(Bieri et al., 2012)	Switzerland	Single	Existing	Opt	N/A
					Sim	Physical model
12	(Boer, 2011)	Austria	Single	Existing	Opt	N/A
					Sim	Delft 3-D
13	(Wan et al., 2010)	China	Single	Existing	Opt	N/A
					Sim	Hydro-dynamic model.
14	(Kummu, et al., 2010)	China-Mekong	Single	Existing	Opt	N/A
					Sim	TE estimation at different level, reservoir, sub-basin and basin levels. Brune method is used.
15	(Bashar et al., 2010)	Blue Nile	Single	Existing	Opt	N/A
					Sim	TE using Brune, and other formula. Note: Siyam formula mentioned here $[e^{-B*V/I}]$ is different from that mentioned by Leo C.van Rijn $[e^{-B*I/V}]$
16	(Minear & Kondolf, 2009)	California-USA	Multiple	Combined	Opt	N/A
					Sim	Estimate Reservoir Sedimentation Rate at large special and temporal scale for existing and planned dams. Used TE based on Brown method.
17	(Khan & Tingsanchali, 2009)	Pakistan	Single	Existing	Opt	Max Sediment evacuated, Power generated, Irrigation supply. Used GA
					Sim	Flushing using Tsinghau University.
18	(Garg & Jothiprakash, 2009)	India	Single	Existing	Opt	N/A
					Sim	TE based on Brune and calibrated to the existing dam
19		China	Single	Existing	Opt	Max Hp and min Sed

No	Study	Country/Region/Basin	Single/Multiple reservoirs	Existing/Planned/Combined	Optimization/Simulation	System condition
	(H. Li & Lian, 2008)				Sim	Mentioned as function of all reservoirs and sediment characteristics. No definite set of equations are shown for Sediment
20	(Dhar & Datta, 2008)	USA	Single	Existing	Opt	Single Objective function: to Min the sum of normalised square deviation of actual storage from a target storage,
					Sim	A hydro-dynamic and sediment transport model is used.
21	(Chang, et al., 2003)	Taiwan	Single	Existing	Opt	Min the number of deficit as results of selecting flushing based operation
					Sim	Tsinghau University formula
22	(J. W. Nicklow & Mays, 2001)	USA	Multiple	Existing	Opt	Successive approximation linear quadratic regulator (SALQR) optimization algorithm to minimise the change of bed level.
					Sim	Sediment transport dynamics
23	(Bringer & Nicklow, 2001)	USA	Multiple	Existing	Opt	GA, mainly for reducing sediment aggregation along channels and reservoirs
					Sim	HEC-6 sediment transport simulation model

REFERENCES

Abate, Z. (1994). The integrated development of Nile waters. In P.P.Howell & J.A.Allan (Eds.), The Nile, sharing a scarce resource, a historical and technical review of water managment and economic and legal issues (pp. 227-242). Cambridge: Cambridge University Press.

Abdallah, M., & Stamm, J. (2013). A Fuzzy Rule Based Operation Model for Blue Nile Reservoirs. Dresden University of Technology. Faculty of Civil Engineering. Institute of Hydraulic Engineering and Technical Hydromechanics, 229-240.

Abreha, Y. G. (2010). Tradeoff between hydropower generation, environmental flow and irrigation: Tekeze river,Nile River Basin, Ethiopia. MSc, UNESCO-IHE Institute for Water Education, Delft.

Afrin, T. (2014). Influence of Flushing Scenarios on Sedimentation in the Reservoir. MSc, UNESCO-IHE Delft. (WES-HI.14-01)

Ahmed, A. A., & Ismail, U. (2008). Sediment in the Nile River system. Consultancy Study requested by UNESCO.

Ali, Y. S. A. (2014). The impact of Soil Erosion in the Upper Blue Nile on Downstream Reservoir Sedimentation. PhD, Delft University of Technology, Delft.

Allan, J. A. (1999). The Nile basin: Evolving approaches to Nile waters management. Occasional Paper, 20, 1-11.

Andjelic, M. (2009). Some features of the NileRiver basin decision support tool. Ingenieurhydrologie und Wasserbewirtschaftung (ihwb), Technische Universität Darmstadt. Retrieved from http://www.ihwb.tu-darmstadt.de/media/fachgebiet_ihwb/lehre/iwrdm/literature/somefeaturesoftheni leriverbasindecisionsupporttool.pdf

Annandale, G. W. (2006). Reservoir Sedimentation Encyclopedia of Hydrological Sciences: John Wiley & Sons, Ltd.

Arjoon, D., Mohamed, Y., Goor, Q., & Tilmant, A. (2014). Hydro-economic risk assessment in the eastern Nile River basin. Water Resources and Economics, 8(Supplement C), 16-31. doi: http://dx.doi.org/10.1016/j.wre.2014.10.004

Arsano, Y., & Tamrat, I. (2005). Ethiopia and the Eastern Nile Basin. Aquatic Sciences - Research Across Boundaries, 67(1), 15-27. doi: 10.1007/s00027-004-0766-x

Asfaw, T. D., & Saiedi, S. (2011). Optimal Short-term Cascade Reservoirs Operation using Genetic Algorithm. Asian Journal of Applied Sciences, 4(3), 297-305. doi: 10.3923/ajaps.2011.297.305

Bai, T., Wu, L., Chang, J.-x., & Huang, Q. (2015). Multi-Objective Optimal Operation Model of Cascade Reservoirs and Its Application on Water and Sediment Regulation. Water Resources Management, 29(8), 2751-2770. doi: 10.1007/s11269-015-0968-0

Barron, M. (2006). A Fact File about the Nile River. In F. R. o. t. Nile (Ed.), 34KB JPG (Vol. 472 × 340): http://www.mbarron.net/Nile/fctfl_nf.html.

Barrow, C. J. (1998). River basin development planning and management: A critical review. [doi: 10.1016/S0305-750X(97)10017-1]. World Development, 26(1), 171-186.

Bashar, K. E., ElTahir, E., Fattah, S. A., Ali, A. S., Musnad, M., & Osman, I. (2010). Nile Basin Reservoir Sedimentation Prediction and Mitigation. Hydraulic Research Institute, Nile Basin Building Capacity Network, Cairo.

Belachew, A., Mekonen, Z., & Ibrahim, Y. (2015). Eastern Nile Basin Water System Simulation Using Hec-ResSim Model. Paper presented at the 11th International Conference on Hydroinformatics, HIC 2014, New York City, USA.

Beyene, T., Lettenmaier, D. P., & Kabat, P. (2010). Hydrologic impacts of climate change on the Nile River Basin: implications of the 2007 IPCC scenarios. Climatic Change, 100(3-4), 433-461.

Bieri, M., Müller, M., Boillat, J., & Schleiss, A. (2012). Modeling of Sediment Management for the Lavey Run-of-River HPP in Switzerland. Journal of Hydraulic Engineering, 138(4), 340-347. doi: doi:10.1061/(ASCE)HY.1943-7900.0000505

Block, P., & Rajagopalan, B. (2007). Interannual Variability and Ensemble Forecast of Upper Blue Nile Basin Kiremt Season Precipitation. [doi: 10.1175/JHM580.1]. Journal of Hydrometeorology, 8(3), 327-343. doi: 10.1175/jhm580.1

Block, P., & Strzepek, K. (2010). Economic Analysis of Large-Scale Upstream River Basin Development on the Blue Nile in Ethiopia Considering Transient Conditions, Climate Variability, and Climate Change. Journal of Water Resources Planning and Management, 136(2), 156-166.

Block, P. J. S., K. Strzepek and B. Rajagopalan. (2007). Integrated management of Blue Nile Basin in Ethiopia under climate variability and change hydropower and irrigation modeling IFPRI discussion papers (Vol. No 700). Washington, DC 20006-1002 USA International Food Policy Research Institute (IFPRI)

Boer, V. d. (2011). Preliminary Study of Flushing Operations of the Langmann Reservoir, Austria. M.S.c, Delft University of Technology, Delft, the Netherlands (1223798)

Borji, T. T. (2013). Sedimentation and Sustainability of Hydropower Reservoirs: Cases of Grand Ethiopian Renaissance Dam on the Blue Nile River in Ethiopia.

Bringer, J. A., & Nicklow, J. W. (2001). Optimal Control of Sedimentation in Multi-Reservoir River Systems Using Genetic Algorithms Bridging the Gap (pp. 1-10).

Brown, C. B. (1958). Sediment transportation. In Rouse, H. (ed.). Engineering hydraulics, Wiley, New York.

Brown, C. M., Lund, J. R., Cai, X., Reed, P. M., Zagona, E. A., Ostfeld, A., Hall, J., Characklis, G. W., Yu, W., & Brekke, L. (2015a). The future of Water Resources Systems analysis: Toward a scientific framework for sustainable water management. Water Resources Research.

Brown, C. M., Lund, J. R., Cai, X., Reed, P. M., Zagona, E. A., Ostfeld, A., Hall, J., Characklis, G. W., Yu, W., & Brekke, L. (2015b). The future of water resources systems analysis: Toward a scientific framework for sustainable water management. Water Resources Research, 51(8), 6110-6124.

Brune, G. M. (1953). Trap efficiency of reservoirs. Trans. AGU, 34(3), 407-418.

Carriaga, C., & Mays, L. (1995a). Optimal Control Approach for Sedimentation Control in Alluvial Rivers. Journal of Water Resources Planning and Management, 121(6), 408-417. doi: doi:10.1061/(ASCE)0733-9496(1995)121:6(408)

Carriaga, C., & Mays, L. (1995b). Optimization Modeling for Sedimentation in Alluvial Rivers. Journal of Water Resources Planning and Management, 121(3), 251-259. doi: doi:10.1061/(ASCE)0733-9496(1995)121:3(251)

Cascão, A. E. (2009). Changing power relations in the Nile River Basin: Unilateralism vs. cooperation. Water Alternatives, 2(2), 245-268.

Celeste, A. B., & Billib, M. (2009). Evaluation of stochastic reservoir operation optimization models. [doi: 10.1016/j.advwatres.2009.06.008]. Advances in Water Resources, 32(9), 1429-1443.

Chang, F.-J., Lai, J.-S., & Kao, L.-S. (2003). Optimization of operation rule curves and flushing schedule in a reservoir. Hydrological Processes, 17(8), 1623-1640. doi: 10.1002/hyp.1204

Chow, V. T., & Cortes-Rivera, G. (1974). Application of DDDP in water resources planning.

Churchill, M. (1948). Discussion of "Analysis and Use of Reservoir Sedimentation Data," by LC Gottschalk. Paper presented at the Proc. of Fedral Interagency Sedimentation Conference, Wahington DC.

Conway, D. (1996). The Impacts of Climate Variability and Future Climate Change in the Nile Basin on Water Resources in Egypt. [doi: 10.1080/07900629650178]. International Journal of Water Resources Development, 12(3), 277-296. doi: 10.1080/07900629650178

Conway, D. (2005). From headwater tributaries to international river: observing and adapting to climate variability and change in the Nile basin. Global Environmental Change, 15(2), 99-114.

Côté, P., & Leconte, R. (2015). Comparison of Stochastic Optimization Algorithms for Hydropower Reservoir Operation with Ensemble Streamflow Prediction. Journal of Water Resources Planning and Management, 04015046.

Dellapenna, J. W. (2001). The Nile as a legal and political structure. Conflict and Cooperation related to International Water Resources: Historical Perspectives, Technical Documents in Hydrology (TDH), 35-48. Retrieved from http://www.hydrologie.org/BIB/Publ_UNESCO/TD_062_2002.pdf#page=39

Demuth, S., & Gandin, J. (2010). FRIEND 2010–Footprints of International Cooperation over a Quarter of a Century. Paper presented at the Proc. of the 6th World FRIEND Conference, Fez, Morocco, IAHS Publ.

Dhar, A., & Datta, B. (2008). Optimal operation of reservoirs for downstream water quality control using linked simulation optimization. Hydrological Processes, 22(6), 842-853. doi: 10.1002/hyp.6651

Di Baldassarre, G., Elshamy, M., van Griensven, A., Soliman, E., Kigobe, M., Ndomba, P., Mutemi, J., Mutua, F., Moges, S., Xuan, Y., Solomatine, D., & Uhlenbrook, S. (2011). Future hydrology and climate in the River Nile basin: a review. Hydrological Sciences Journal, 56(2), 199-211. doi: 10.1080/02626667.2011.557378

Digna, R.F., Mohamed, Y.A., van der Zaag, P., Uhlenbrook, S. and Corzo, G.A. (2018a). Impact of wa-ter resources development on water availability for hydropower production and irrigated agriculture of the Eastern Nile Basin. ASCE Journal of Water Resource Planning and Management 144(5): 05018007

Digna, R., Castro-Gama, M., van der Zaag, P., Mohamed, Y., Corzo, G., & Uhlenbrook, S. (2018b). Optimal Operation of the Eastern Nile System Using Genetic Algorithm, and Benefits Distribution of Water Resources Development. Water, 10(7), 921.

Digna, R. F., Mohamed, Y. A., van der Zaag, P., Uhlenbrook, S., & Corzo, G. A. (2017). Nile River Basin modelling for water resources management – a literature review. International Journal of River Basin Management, 15(1), 39-52. doi: 10.1080/15715124.2016.1228656

Dinar, A., Dinar, S., & McCaffrey, S. (2007). Bridges over water: understanding transboundary water conflict, negotiation and cooperation (Vol. 11): World Scientific.

Dinar, A., & Nigatu, G. S. (2013). Distributional considerations of international water resources under externality: The case of Ethiopia, Sudan and Egypt on the Blue Nile. Water Resources and Economics, 2–3(0), 1-16. doi: http://dx.doi.org/10.1016/j.wre.2013.07.001

Dombrowsky, I. (2009). Revisiting the potential for benefit sharing in the management of trans-boundary rivers. Water Policy 11(2), 125:145. doi: 10.2166/wp.2009.020

El-Fadel, M., El-Sayegh, Y., El-Fadl, K., & Khorbotly, D. (2003). The Nile River Basin: a case study in surface water conflict resolution. Journal of Natural Resources and Life Sciences Education, 32, 107-117.

Elimam, L., Rheinheimer, D., Connell, C., & Madani, K. (2008). An ancient struggle: a game theory approach to resolving the Nile conflict. Paper presented at the Proceeding of the 2008 world environmental and water resources congress. American Society of Civil Engineers. Honolulu, Hawaii.

Elshopky, M. (2012). The Impact of Water Scarcity on Egyptian National Security and on Regional Security in the Nile River Basin: DTIC Document.

Eltahir, E. A. B. (1996). El Niño and the Natural Variability in the Flow of the Nile River. Water Resources Research, 32(1), 131-137. doi: 10.1029/95wr02968

ENTRO. (2007). Multipurpose Development of the Eastern Nile, One-System inventorySynthesis work Report. In M. Amare (Ed.). Addis Ababa: Eastern Nile Technical Regional Office.

Fahmi, A. M. (2007). Water Management in the Nile Basin, Opportunitites and Constraints. no 143.

Garcâia, M. H. (2008). Sedimentation engineering: processes, measurements, modeling, and practice: ASCE Publications.

Garg, V., & Jothiprakash, V. (2009). Estimation of useful life of a reservoir using sediment trap efficiency. Journal of Spatial Hydrology, 8(2).

Georgakakos, A., Yao, H., Brumbelow, K., DeMarchi, C., Bourne, S., and Mullusky, M. (2000). The Lake Victoria decision support system, prepared for the Food and Agriculture Organization of the United Nations: Technical Report No. GWRI-2000-1, Georgia Water Resources Institute and Georgia Tech.

Georgakakos, A. P. (2006). Decision support systems for water resources management: State of the science review and Nile basin applications Paper presented at the IN: Proceedings of the International Working Conference: Enhancing Equitable Livelihood Benefits of Dams Using Decision Support Systems, Nazareth, Ethiopia, 23-26 January 2006. http://hdl.handle.net/10568/21474

Georgakakos, A. P. (2007). Decision support systems for integrated water resources management with an application to the Nile basin. Topics on system analysis and integrated water resource management.

Ghany, H. E. A. (1994). Application of Stochastic-Dynamic Programming for the Operation of a Two- Reservoir on The Blue Nile in the Sudan. MSc Thesis, International Institute for Hydraulic and Environmental Engineering, Delft.

Giordano, R., Passarella, G., Uricchio, V., & Vurro, M. (2005). Fuzzy cognitive maps for issue identification in a water resources conflict resolution system. Physics and Chemistry of the Earth, Parts A/B/C, 30(6), 463-469.

Gismalla, Y. (2009). Sedimentation Problems in the Blue Nile Reservoirs and Gezira Scheme: A Review. Gezira Journal of Engineering and Applied Science, 14(2), 1-12.

González Iñiguez, A. A. (2017). Modelo para la optimización de operaciones en embaldes con evacuación de sedimentos. Caso de estudio: Embalse Manduriacu-Ecuador. Bélgica: Universiteit Katholieke.

Goor, Q., Halleux, C., Mohamed, Y., & Tilmant, A. (2010). Optimal operation of a multipurpose multireservoir system in the Eastern Nile River Basin. Hydrol. Earth Syst. Sci., 14(10), 1895: 1908 doi: 10.5194/hess-14-1985-2010

Greg Schellenberg, C. Richard Donnelly, Charles Holder, M.-H. B., & Rajib Ahsan, H. (2017). Sedimentation, Dam Safety and Hydropower: Issues, Impacts and Solutions. Sedimentation and Hydropower: Impacts and Solutions. Retrieved from https://www.hydroworld.com/library/2017/04/Sedimentation-Dam-Safety-and-Hydropower-Issues-Impacts-and-Solutions.html website:

Griensven, A. van, Ndomba, P., Yalew, S., & Kilonzo, F. (2012). Critical review of SWAT applications in the upper Nile basin countries. Hydrology and Earth System Sciences, 16(9), 3371-3381.

Guariso, G., Haynes, K. E., Whittington, D., & Younis, M. (1981). A Real-Time Management Model for the Aswan High Dam with Policy Implications. Geographical Analysis, 13(4), 355-372. doi: 10.1111/j.1538-4632.1981.tb00744.x

Guariso, G., & Whittington, D. (1987). Implications of ethiopian water development for Egypt and Sudan. [doi: 10.1080/07900628708722338]. International Journal of Water Resources Development, 3(2), 105-114. doi: 10.1080/07900628708722338

Habteyes, B. G., Hasseen El-bardisy, H. A. E., Amer, S. A., Schneider, V. R., & Ward, F. A. (2015). Mutually beneficial and sustainable management of Ethiopian and Egyptian dams in the Nile Basin. Journal of Hydrology, 529, Part 3, 1235-1246. doi: http://dx.doi.org/10.1016/j.jhydrol.2015.09.017

Hajiabadi, R., & Zarghami, M. (2014). Multi-objective reservoir operation with sediment flushing; case study of Sefidrud reservoir. Water Resources Management, 28(15), 5357-5376.

Hakimi-Asiabar, M., Ghodsypour, S. H., & Kerachian, R. (2010). Deriving operating policies for multi-objective reservoir systems: Application of Self-Learning Genetic Algorithm. [doi: 10.1016/j.asoc.2009.08.016]. Applied Soft Computing, 10(4), 1151-1163.

Hamad, O. E. (1993). Optimal Operation of A Reservoir System During A Dry Season. PhD Thesis, University of NewCastle Upon Tyne.

Hamid, S. H. (2013). The impacts of planned Ethiopian dams on the Blue Nile System in Sudan using the NBDSS   Paper   presented   at   the New Nile Perspectives conference Khartoum.

Hammond, M. (2013). The Grand Ethiopian Renaissance Dam and the Blue Nile: Implications for transboundary water governance. Paper presented at the Global Water Forum.

Harou, J. J., Pulido-Velazquez, M., Rosenberg, D. E., Medellín-Azuara, J., Lund, J. R., & Howitt, R. E. (2009). Hydro-economic models: Concepts, design, applications, and future prospects. [doi: 10.1016/j.jhydrol.2009.06.037]. Journal of Hydrology, 375(3–4), 627-643.

Hassaballah, K. (2010). Model-Based Optimization of Downstream Impact during Filling of a New Reservoir: Case Study of Mandaya/Roseires Reservoirs on the Blue Nile River. MSc, UNESCO-IHE Institute for Water Education, Delft. Retrieved from http://dx.doi.org/10.1007/s11269-011-9917-8

Hassaballah, K., Jonoski, A., Popescu, I., & Solomatine, D. (2011). Model-Based Optimization of Downstream Impact during Filling of a New Reservoir: Case Study of Mandaya/Roseires Reservoirs on the Blue Nile River. Water Resources Management, 1-21. doi: 10.1007/s11269-011-9917-8

Hipel, K., Kilgour, D., Fang, L., & Li, K. (2002). Resolution of water conflicts between Canada and the United States. Paper submitted for the PCCP Project report.

Hurst, H. E., Black, R. P., & Simaika, Y. M. (1966). The Nile Basin, Vol. X: The Main Nile Projects. Cairo: General Organization for Government Printing Office.

Jacoby, H. D., & Loucks, D. P. (1972). Combined use of optimization and simulation models in river basin planning. Water Resour. Res., 8(6), 1401-1414. doi: 10.1029/WR008i006p01401.

Jeuland, M. (2010). Economic implications of climate change for infrastructure planning in transboundary water systems: An example from the Blue Nile. Water Resources Research, 46(11).

Jeuland, M., & Whittington, D. (2014). Water resources planning under climate change: Assessing the robustness of real options for the Blue Nile. Water Resources Research, 50(3), 2086-2107.

Jeuland, M., Wu, X., & Whittington, D. (2017). Infrastructure development and the economics of cooperation in the Eastern Nile. Water International, 42(2), 121-141. doi: 10.1080/02508060.2017.1278577.

Kahsay, T. N. (2017). Towards Sustainable Water Resources Management in the Nile River Basin. A Global Computable General Equilibrium Analysis. PhD, Vrije University Amsterdam.

Kahsay, T. N., Kuik, O., Brouwer, R., & van der Zaag, P. (2015). Estimation of the transboundary economic impacts of the Grand Ethiopia Renaissance Dam: A computable general equilibrium analysis. Water Resources and Economics, 10, 14-30.

Karamouz, M., Akhbari, M., & Moridi, A. (2011). Resolving Disputes over Reservoir-River Operation. Journal of Irrigation and Drainage Engineering, 137(5), 327-339.

Karamouz, M., Szidarovszky, F. (2003). Water Resources Systems analysis: with Emphasisnon Conflict Resolution Retrieved from http://xa.yimg.com/kq/groups/21948400/1183711781/name/Water_Resources_S ystems_Analysis_by_Karamouz.pdf

Karyabwite, D. R. (2000). Water sharing in the Nile River valley. Project GNV011: Using GIS/Remote Sensing for the sustainable use of natural resources. United Nations Environment Programme, Nairobi.

Kelman, J., Stedinger, J. R., Cooper, L., Hsu, E., & Yuan, S. Q. (1990). Sampling stochastic dynamic programming applied to reservoir operation. Water Resources Research, 26(3), 447-454.

Khan, N. M., & Tingsanchali, T. (2009). Optimization and simulation of reservoir operation with sediment evacuation: a case study of the Tarbela Dam, Pakistan. Hydrological Processes, 23(5), 730-747. doi: 10.1002/hyp.7173

Kim, T., Heo, J.-H., & Jeong, C.-S. (2006). Multireservoir system optimization in the Han River basin using multi-objective genetic algorithms. Hydrological Processes, 20(9), 2057-2075. doi: 10.1002/hyp.6047

Kim, U., Kaluarachchi, J. J., & Smakhtin, V. U. (2008). Climate change impacts on hydrology and water resources of the Upper Blue Nile River Basin, Ethiopia (Vol. 126, pp. 21). Colombo, Sri Lanka: International Water Management Institute (IWMI).

King, A., & Block, P. (2014). An assessment of reservoir filling policies for the Grand Ethiopian Renaissance Dam. Journal of Water and Climate Change, 5(2), 233-243.

Kondolf, G. (2013). Sustainable sediment management in reservoirs and regulated rivers: Experiences from five continents, Earth's Future, 2, doi: 10.1002/2013EF000184: Received.

Kougias, I. P., & Theodossiou, N. P. (2013). Application of the harmony search optimization algorithm for the solution of the multiple dam system scheduling. Optimization and Engineering, 14(2), 331-344.

Kucukmehmetoglu, M. (2009). A Game Theoretic Approach to Assess the Impacts of Major Investments on Transboundary Water Resources: The Case of the Euphrates and Tigris. Water Resources Management, 23(15), 3069-3099. doi: 10.1007/s11269-009-9424-3.

Kummu, M., Lu, X. X., Wang, J. J., & Varis, O. (2010). Basin-wide sediment trapping efficiency of emerging reservoirs along the Mekong. [doi: 10.1016/j.geomorph.2010.03.018]. Geomorphology, 119(3–4), 181-197.

Labadie, J. W. (2004). Optimal Operation of Multireservoir Systems: State-of-the-Art Review. Journal of Water Resources Planning and Management, 130(2), 93-111.

Larijani, K. M. (2009). climate change effects on high-elevation hydropower system in California. UNIVERSITY OF CALIFORNIA.

Larson, R. E. (1968). State increment dynamic programming.

Lee, C., & Foster, G. (2013). Assessing the potential of reservoir outflow management to reduce sedimentation using continuous turbidity monitoring and reservoir modelling. Hydrological Processes, 27(10), 1426-1439.

Lee, Y., Yoon, T., & Shah, F. A. (2012). Optimal Watershed Management for Reservoir Sustainability: Economic Appraisal. Journal of Water Resources Planning and Management, 139(2), 129-138; doi:10.1061/(ASCE)WR.1943-5452.0000232.

Levy, B. S., & Baecher, G. B. (1999). NileSim: A Windows-based hydrologic simulator of the Nile River Basin. Journal of Water Resources Planning and Management, 125(2), 100-106.

Lewis, S. E., Bainbridge, Z. T., Kuhnert, P. M., Sherman, B. S., Henderson, B., Dougall, C., Cooper, M., & Brodie, J. E. (2013). Calculating sediment trapping efficiencies for reservoirs in tropical settings: a case study from the Burdekin Falls Dam, NE Australia. Water Resources Research, 49(2), 1017-1029.

Li, F., Shoemaker, C., Wei, J., & Fu, X. (2013). Estimating Maximal Annual Energy Given Heterogeneous Hydropower Generating Units with Application to the Three Gorges System. Journal of Water Resources Planning and Management, 139(3), 265-276. doi: doi:10.1061/(ASCE)WR.1943-5452.0000250.

Li, H., & Lian, J. (2008). Multi-objective optimization of water-sedimentation-power in reservoir based on pareto-optimal solution. Transactions of Tianjin University, 14(4), 282-288. doi: 10.1007/s12209-008-0048-0

Loucks, D. P. (1979). Water resources systems. Rev. Geophys., 17(6), 1335-1351. doi: 10.1029/RG017i006p01335.

Loucks, D. P., & Van Beek, E. (2005). Water Resources Systems Planning and Management.

An Introduction to Methods, Models, and Applications: United Nations Educational, Sceintific and Cultural Organization.

Lund, J. R. G., J. (1999). Some Derived Operating Rules for Reservoirs in Series or in Parallel. Journal of Water Resources Planning and Management, 125(3), 143-153.

Madani, K. (2010). Game theory and water resources. [doi: 10.1016/j.jhydrol.2009.11.045]. Journal of Hydrology, 381(3–4), 225-238.

Mastrandrea, M. D., Mach, K. J., Plattner, G.-K., Edenhofer, O., Stocker, T. F., Field, C. B., Ebi, K. L., & Matschoss, P. R. (2011). The IPCC AR5 guidance note on consistent treatment of uncertainties: a common approach across the working groups. Climatic Change, 108(4), 675-691.

McCartney, M. P. (2007). Decision support systems for large dam planning and operation in Africa (Vol. 119): IWMI.

McCartney, M. P., & Menker Girma, M. (2012). Evaluating the downstream implications of planned water resource development in the Ethiopian portion of the Blue Nile River. Water International, 37(4), 362-379. doi: 10.1080/02508060.2012.706384

McKinney, D. C., Cai, X., Rosegrant, M. W., Ringler, C., & Scott, C. A. (1999). Modeling Water Resources Management at the Basin Level: Review and Future Directions (Vol. SWIM Paper 6): IWMI.

McLellan, M. a. (1987). Updating of the Feasibility Study for the Heightening of Roseires Dam. Final Report (Vol. 2- Appendices): SIR ALEXANDER GIBB & PARTNERS.

Mekonnen, D. Z. (2010). The Nile basin cooperative framework agreement negotiations and the adoption of a 'Water Security'paradigm: Flight into obscurity or a logical cul-de-sac? European Journal of International Law, 21(2), 421-440.

Metawie, A. F., & Sector, N. W. (2004). Lessons Learnt From Cooperation in the Nile Basin. Paper presented at the Fourth Biennial Rosenberg International Forum on Water Policy. Retrieved from http://rosenberg. ucanr. org/forum4. cfm.

Minear, J. T., & Kondolf, G. M. (2009). Estimating reservoir sedimentation rates at large spatial and temporal scales: A case study of California. Water Resour. Res., 45(12), W12502. doi: 10.1029/2007wr006703.

Ministry of Agriculture. (2013). Timeseries of Area Planted, Harvested, Production & Yield Data of the Main Food & Oil Crops by Production Centers & Type of Irrigation; Ministry of Agriculture: Khartoum, Sudan.

Mirchi, A., Waktins Jr, D., & Madani, K. (2010). Modeling for watershed planning, management, and decision making. Watersheds: management, restoration, and environmental impact. New York: Nova Science Publishers.

Mohamed, Y. A. (1990). Simulation and optimization of the Blue Nile double reservoir system. MSc Thesis, International Institute for Hydraulic and Environmental Engineering, Delft.

Mohamed, Y. A. (2011). The hydrological study of the Baro Akobo Sobat Basin Baro Akobo Sobat Project: Eaastern Nile Technical Regional Office(ENTRO).

MOI. (1979). Nile Waters Master Plan Summary. Consultants: Coyne Et bellier, Sir Alexander Gibb and Partners, Hunting Technical Services Limited, Sir M Macdonalds and Partners.

Momtahen, S., & Dariane, A. (2007). Direct Search Approaches Using Genetic Algorithms for Optimization of Water Reservoir Operating Policies. Journal of Water Resources Planning and Management, 133(3), 202-209. doi: doi:10.1061/(ASCE)0733-9496(2007)133:3(202).

Moriasi, D., Arnold, J., Van Liew, M., Bingner, R., Harmel, R., & Veith, T. (2007). Model evaluation guidelines for systematic quantification of accuracy in watershed simulations. Transactions of the ASABE, 50(3), 885-900.

Morrice, H. A. W., & Allan, W. N. (1958). Report On the Nile Valley Plan. Volume(1) Volume(1) (Vol. Volume(1)): Ministry of Water Resources and Electricity, Sudan.

Morris, G. L., & Fan, J. (1998). Reservoir sedimentation handbook: design and management of dams, reservoirs, and watersheds for sustainable use (Vol. 9): McGraw-Hill New York.

Mulat, A. G., & Moges, S. A. (2014a). Assessment of the Impact of the Grand Ethiopian Renaissance Dam on the Performance of the High Aswan Dam. Journal of Water Resource and Protection, No.06, 583-598. doi: 10.4236/jwarp.2014.66057

Mulat, A. G., & Moges, S. A. (2014b). Filling Option Assessments for Proposed Reservoirs in Abbay (Upper Blue Nile) River Basin to Minimize Impacts on Energy Generation of Downstream Reservoirs. Open Journal of Renewable Energy and Sustainable Development, 1(1), 14.

Murray, D. M., & Yakowitz, S. J. (1979). Constrained differential dynamic programming and its application to multireservoir control. Water Resour. Res., 15(5), 1017-1027. doi: 10.1029/WR015i005p01017

Musa, Y. A. (1985). Development of River Basin Operational Guidlines For Favorable Distribution of Shortages. PhD Thesis, Colardo State University, Fort Collins, Colorado.

NBI. (2010). The Cooperative Framework Agreement for the River Nile Basin: An Overview Agreement on the Nile River Basin Cooperative Framework from http://www.internationalwaterlaw.org/documents/regionaldocs/Nile_River_Basin_Cooperative_Framework_2010.pdf

NBI. (2012). The State of the River Nile Basin 2012. Entebbe, : Nile Basin Initiative.

Nicklow, J., & Mays, L. (2000). Optimization of Multiple Reservoir Networks for Sedimentation Control. Journal of Hydraulic Engineering, 126(4), 232-242. doi: doi:10.1061/(ASCE)0733-9429(2000)126:4(232)

Nicklow, J., Reed, P., Savic, D., Dessalegne, T., Harrell, L., Chan-Hilton, A., Karamouz, M., Minsker, B., Ostfeld, A., Singh, A., Environmental, E. Z. A. T. C. o. E. C. i., & Engineering, W. R. (2010). State of the Art for Genetic Algorithms and Beyond in Water Resources Planning and Management. Journal of Water Resources Planning and Management, 136(4), 412-432.

Nicklow, J. W., & Mays, L. W. (2001). Optimal Control of Reservoir Releases to Minimize Sedimentation in Rivers and Reservoirs. JAWRA Journal of the American Water Resources Association, 37(1), 197-211. doi: 10.1111/j.1752-1688.2001.tb05486.x

Osman, I. (2015). Impact of improved operation and maintenance on cohesive sediment transport in Gezira Scheme, Sudan. PhD, Wageningen University and UNESCO-IHE Institute for Water Education, Delft.

Oven-Thompson, K., Alercon, L., & Marks, D. H. (1982). Agricultural vs. hydropower tradeoffs in the operation of the High Aswan Dam. Water Resources Research, 18(6), 1605-1613.

Pannell, D. J. (1997). Sensitivity analysis: strategies, methods, concepts, examples. Agric Econ, 16, 139-152.

Philbrick, C. R., & Kitanidis, P. K. (1999). Limitations of deterministic optimization applied to reservoir operations. Journal of Water Resources Planning and Management, 125(3), 135-142.

Pokharel, S. (2007). Water use opportunities and conflicts in a small watershed—a case study. Renewable and Sustainable Energy Reviews, 11(6), 1288-1299.

Qin, Z., Zhang, J., DiTommaso, A., Wang, R., & Liang, K. (2016). Predicting the potential distribution of Lantana camara L. under RCP scenarios using ISI-MIP models. Climatic Change, 134(1-2), 193-208.

Rani, D., & Moreira, M. (2010). Simulation–Optimization Modeling: A Survey and Potential Application in Reservoir Systems Operation. Water Resources Management, 24(6), 1107-1138. doi: 10.1007/s11269-009-9488-0.

Rashid, M. U., Shakir, A. S., Khan, N. M., Latif, A., & Qureshi, M. M. (2015). Optimization of Multiple Reservoirs Operation with Consideration to Sediment Evacuation. [journal article]. Water Resources Management, 29(7), 2429-2450. doi: 10.1007/s11269-015-0951-9.

Ribbe, L., & Ahmed, S. (2006). Transboundary Water Management in the Nile River Basin. Technology Resource Management and Development 4, 13-27. doi: http://www.tt.th-koeln.de/publications/http://www.tt.fh-koeln.de/publications/ittpub%20303101_03.pdf

Riediger, J., Breckling, B., Svoboda, N., & Schröder, W. (2016). Modelling regional variability of irrigation requirements due to climate change in Northern Germany. Science of The Total Environment, 541, 329-340. doi: http://dx.doi.org/10.1016/j.scitotenv.2015.09.043.

Rogelj, J., Meinshausen, M., & Knutti, R. (2012). Global warming under old and new scenarios using IPCC climate sensitivity range estimates. Nature Climate Change, 2(4), 248-253.

Rogers, P. (1969). A game theory approach to the problems of international river basins. Water Resources Research, 5(4), 749-760.

Ryu, J. H., Palmer, R. N., Jeong, S., Lee, J. H., & Kim, Y. O. (2009). Sustainable Water Resources Management in a Conflict Resolution Framework1. JAWRA Journal of the American Water Resources Association, 45(2), 485-499.

Sadoff, C. W., & Grey, D. (2002). Beyond the river: the benefits of cooperation on international rivers. Water Policy, 4(5), 389-403.

Salame, L., & Van der Zaag, P. (2010). Enhanced knowledge and education systems for strengthening the capacity of transboundary water management. In A. J. Anton Earle, Joakim Ojendal (Ed.), Transboundary Water Management: Principles and Practice (pp. 171-186). Earthscan, London.

Salman, S. M. (2013). The Nile Basin Cooperative Framework Agreement: a peacefully unfolding African spring? Water International, 38(1), 17-29.

Salman, S. M. A. (2016). The Grand Ethiopian Renaissance Dam: the road to the declaration of principles and the Khartoum document. Water International, 41(4), 512-527. doi: 10.1080/02508060.2016.1170374.

Samaan, M. M. (2014). The Win-Win-Win Scenario in the Blue Nile's Hydropolitical Game: Application on the Grand Ethiopian Renaissance Dam.

Satti, S., Zaitchik, B., & Siddiqui, S. (2014). The question of Sudan: a hydroeconomic optimization model for the Sudanese Nile. Hydrology and Earth System Sciences Discussions, 11(10), 11565-11603.

Savenije, H. H. (1995). Spreadsheets: Flexible Tools for Integrated Management of Water Resources in River Basins. In: Modelling and Management of Sustainable Basin-scale Water Resources Systems. IAHS Publications no. 231, 207-215.

Sayed, D. M. A. A. (2008). Eastern Nile Planning Model, Integration with IDEN Projects to Deal with Climate Change Uncertainty and Flooding Risk. Nile Water Science & Engineering Journal, 1(1).

Schleiss, A. J., Franca, M. J., Juez, C., & De Cesare, G. (2016). Reservoir sedimentation. Journal of Hydraulic Research, 54(6), 595-614. doi: 10.1080/00221686.2016.1225320.

Sechi, G. M., & Sulis, A. (2009). Water System Management through a Mixed Optimization-Simulation Approach. Journal of Water Resources Planning and Management, 135(3), 160-170.

Shahin, M. (1985). Hydrology of the Nile basin (illustrated ed. Vol. Volume 21 of Developments in water science). Amsterdam, The Netherlands: Access Online via Elsevier.

Shokri, A., Haddad, O., & Mariño, M. (2013). Reservoir Operation for Simultaneously Meeting Water Demand and Sediment Flushing: Stochastic Dynamic Programming Approach with Two Uncertainties. Journal of Water Resources Planning and Management, 139(3), 277-289. doi:10.1061/(ASCE)WR.1943-5452.0000244.

Siam, M. S., & Eltahir, E. A. B. (2015). Explaining and forecasting interannual variability in the flow of the Nile River. Hydrol. Earth Syst. Sci., 19(3), 1181-1192. doi: 10.5194/hess-19-1181-2015.

Simonovic, S. P. (1992). Reservoir Systems Analysis: Closing Gap between Theory and Practice. Journal of Water Resources Planning and Management, 118(3), 262-280.

Siyam, A., Yeoh, J., & Loveless, J. (2001). Sustainable reservoir sedimentation control. Paper presented at the Proceedings of the Congress-International Association for Hydraulic Research.

Siyam, A. M., Mirghani, M., Elzein, S., Golla, S., & El-sayed, S. (2005). Assessment of the current state of the Nile basin reservoir sedimentation problems. Nile Basin Capacity Building Network (NBCBN), River morphology Research Cluster, Group1.

Sloff, C. (1991). Reservoir Sedimentation: A literature Survey (Vol. 91-2). Delft: Faculty of Civil Engineering, Delft University of Technology.

Stedinger, J. R., Sule, B. F., & Loucks, D. P. (1984). Stochastic dynamic programming models for reservoir operation optimization. Water Resour. Res., 20(11), 1499-1505. doi: 10.1029/WR020i011p01499

Steduto, P., Hsiao, T. C., Raes, D., & Fereres, E. (2012). Crop yield response to water: Food and Agriculture Organization of the United Nations Rome.

Subramanian, A., Brown, B., & Wolf, A. (2012). Reaching across the waters: Facing the Risks of Cooperation in International waters. Washington DC: World Bank Publications.

Sutcliffe, J. V., & Parks, Y. P. (1999). The hydrology of the Nile. Wallingford, Oxfordshire, UK: International Association of Hydrological Sciences (IAHS).

Taye, M. T., Ntegeka, V., Ogiramoi, N. P., & Willems, P. (2011). Assessment of climate change impact on hydrological extremes in two source regions of the Nile River Basin. Hydrol. Earth Syst. Sci., 15(1), 209-222. doi: 10.5194/hess-15-209-2011

Taye, M. T., & Willems, P. (2012). Temporal variability of hydroclimatic extremes in the Blue Nile basin. [W03513]. Water Resources Research, 48(3).

Taye, M. T., Willems, P., & Block, P. (2015). Implications of climate change on hydrological extremes in the Blue Nile basin: A review. Journal of Hydrology: Regional Studies, 4(Part B), 280-293.

Tebbi, F. Z., Dridi, H., & Morris, G. L. (2012). Optimization of cumulative trapped sediment curve for an arid zone reservoir: Foum El Kherza (Biskra, Algeria). [doi: 10.1080/02626667.2012.712740]. Hydrological Sciences Journal, 57(7), 1368-1377. doi: 10.1080/02626667.2012.712740

Timmerman, J. G. (2005). Transboundary river basin management regimes: the Nile basin case study (Vol.): Background report to Deliverable 1.3.1. of the NeWater project, Lelystad.

Van der Krogt, W. N. M. (2008). RIBASIM Version 7 Technical Reference Manual Deltares.

Van der Krogt, W. N. M., & Boccalon, A. (2013). River Basin Simulation Model RIBASIM Version 7.00 User Manual. Deltares.

Van der Krogt, W. N. M., & Ogink, I. H. J. M. (2013). Development of the Eastern Nile Water Simulation Model, Main Report (Vol. 1206020-000-VEB-0010): Deltares.

Verhaeghe, R. J., Krogt, W. v. d., & Most, H. v. d. (1988). Simulation and Optimization Analysis of Water Resources of Tana River Basin in Kenya. Paper presented at the 6th Conngres Asian and Pacific Regional Division of International Association for Hydraulic Research, Keyota, Japan.

Verhoeven, H. (2011). Black Gold for Blue Gold? Sudan's Oil, Ethiopia's Water and Regional Integration Chatham House Briefing Papers (Vol. 3, pp. 24): The Royal Institute of International Affairs.

Wan, X., Wang, G., Yi, P., & Bao, W. (2010). Similarity-Based Optimal Operation of Water and Sediment in a Sediment-Laden Reservoir. Water Resources Management, 24(15), 4381-4402. doi: 10.1007/s11269-010-9664-2

Wardlaw, R., & Sharif, M. (1999). Evaluation of genetic algorithms for optimal reservoir system operation. Journal of Water Resources Planning and Management, 125(1), 25-33.

Wassie, Y. A. (2008). Decision Support System for Lake Tana Basin, Ethiopia. MSC, UNESCO-IHE institute for Water Education, Dleft.

Wheeler, K. G., Basheer, M., Mekonnen, Z. T., Eltoum, S. O., Mersha, A., Abdo, G. M., Zagona, E. A., Hall, J. W., & Dadson, S. J. (2016). Cooperative filling approaches for the Grand Ethiopian Renaissance Dam. Water International, 1-24.

Whittington, D. (2004). Visions of Nile basin development. Water Policy 2004, 6, 1-24.

Whittington, D., Wu, X., & Sadoff, C. (2005). Water resources management in the Nile basin: the economic value of cooperation. Water Policy, 7, 227-252.

Wolf, A., & Newton, J. (2013). Case study of transboundary dispute resolution: the Nile Waters Agreement. Institute for Water and Watersheds.

Wondimagegnehu, D., & Tadele, K. (2015). Evaluation of climate change impact on Blue Nile Basin Cascade Reservoir operation-case study of proposed reservoirs in the Main Blue Nile River Basin, Ethiopia. Proceedings of the International Association of Hydrological Sciences, 366, 133-133.

Wu, X., & Whittington, D. (2006). Incentive compatibility and conflict resolution in international river basins: A case study of the Nile Basin. Water Resources Research, 42(2), W02417.

Wurbs, R. (1991). Optimization of Multiple-Purpose Reservoir System Operations: A Review of Modeling and Analysis Approaches (Vol. RD-34): US Army Corps of Engineers, Hydrologic Engineering Center.

Wurbs, R. A. (1993). Reservoir-system simulation and optimization models. Journal of Water Resources Planning and Management, 119(4), 455-472.

Yeh, W. W. G. (1985). Reservoir Management and Operations Models: A State-of-the-Art Review. Water Resour. Res., 21(12), 1797-1818. doi: 10.1029/WR021i012p01797

Young, H., Okada, N., & Hashimoto, T. (1980). Cost allocation in water resources development-a case study of Sweden. Research Report, International Institute for Applied Systems Analysis(RR-80-32).

Zaroug, M. A. H., Eltahir, E. A. B., & Giorgi, F. (2014). Droughts and floods over the upper catchment of the Blue Nile and their connections to the timing of El Niño and La Niña events. Hydrol. Earth Syst. Sci., 18(3), 1239-1249. doi: 10.5194/hess-18-1239-2014.

Zhang, Y., Erkyihum, S. T., & Block, P. (2016). Filling the GERD: evaluating hydroclimatic variability and impoundment strategies for Blue Nile riparian countries. Water International, 41 (4), 593-610. doi: 10.1080/02508060.2016.1178467.

LIST OF ACRONYMS

AHD	Aswan High Dam
ANN	Artificial Neural Network
ASO	Ant Colony Optimization
CCLP	Chance Constrained Linear Programming
CDDP	Constrained Differential Dynamic Programming
CI	Computational Intelligence
DDP	Differential Dynamic Programming
DDDP	Discrete Differential Dynamic Programming
DLP	Deterministic Linear Programming
DPSA	Dynamic Programming with Successive Approximation
DSM	Decision Support Model
DST	Decision Support Tool
EA	Evolutionary Algorithm
EC	Evolutionary Computation
EN	Eastern Nile
ENTRO	Eastern Nile Regional Office
ESO	Explicit Stochastic Optimization
FDP	Folded Dynamic Programming
GA	Genetic Algorithm
GERD	Grand Ethiopian Renaissance Dam

HBMO	Honey Bees Mating Optimization
IDP	Incremental Dynamic Programming
IDPSA	Incremental Dynamic Programming with Successive Approximation
ISO	Implicit Stochastic Optimization
IWROM	Integrated Water Resources Optimization Modelling
JMP	Joint Multipurpose Projects
LP	Linear Programming
m.a.s.l.	Meter Above Sea Level
MOGA	Multi Objective Genetic Algorithm
MOO	Multi Objective Optimization
NBI	Nile Basin Initiatives
NDP	Neuro Dynamic Programming
NLP	Non Linear Programming
PSO	Particle Swarm Optimization
SA	Simulated Annealing
SAP	Subsidiary Action Program
SDP	Stochastic Dynamic Programming
SDDP	Stochastic Dual Dynamic Programming
SENSE	Research School for (Socio - Economic and Natural Sciences of Environment)
SLP	Stochastic Linear Programming
SSDP	Sampling Stochastic Dynamic Programming

SVP	Shared Vision Program
TE	Trap Efficiency
TS	Tabu Search
USBR	United State of Bureau of Reclamation
WSE	Water Science and Engineering

LIST OF SYMBOLS

Symbol	Parameter description	Value/ unit
A_i	Irrigated area of scheme (i)	m^2
A_j	Drainage area of reservoir(j)	Km^2
A_{oj}	Surface area of reservoir (j) at the dead storage level	m^2
A_{tj}	The area per unit storage of reservoir (j)	m^2/m^3
C	Constant represents specific gravity and unit conversion	N/m^3
Csd	Trap efficiency coefficient	-
$C_{j,k}^R$	Reservoir system connectivity matrix = −1 when abstraction, +1, receives water from upstream reservoir [reservoir (j) receives water from reservoir (K)]	-
$C_{j,z}^{IR}$	Irrigation system connectivity matrix = −1 when abstraction, +1, receives return water from upstream irrigation [reservoir (j) receives water from irrigation (i)]	-
$CW_{t,i}$	Crop water requirement of irrigation scheme (i) at time (t)	m/month
c_1, c_2, c_3	Constants, representing the weight of the penalty terms in the objective function	
D_j	Target end storage of reservoir (j) at time (T)	m^3
DSD_j	Required flow downstream Roseires Reservoir, including environmental flow and downstream demand at Khartoum, estimated at 244 x 10^6 m^3/month	m^3/month
Ev_t	Monthly Evaporation rate from unit area of surface	m/month
$e_{t,j}$	Evaporation loss of reservoir (j) at time (t)	m^3/month
F	Total Objective function	US\$/m^3

Symbol	Parameter description	Value/ unit
f1	Objective function-1, Hydropower generation	US$
f2	Objective function-2, Irrigation supply	US$/m^3
f3	Objective function-3, Sediment release	US$/m^3
$FIR_{t,j}$	Flushing operation	-
$H^{net}_{t,j}$	Turbine Net Head of reservoir (j) at time (t)	m
$HP_{t,j}$	Total generated energy from Reservoir (j) at time (t)	MWh/month
$HP^{max}_{t,j}$	Maximum hydropower energy could be generated from reservoir (j) at time (t)	MWh/month
I	Total number of irrigation schemes in the system	-
I_t	Inflow state variables at time (t)	m^3/month
$IR_{t,i}$	Withdrawn water for irrigation (i) at time (t)	m^3/month
$IR^{min}_{t,i}$	Minimum water withdrawn for irrigation (i) at time (t)	m^3
$IR^{max}_{t,i}$	Maximum water withdrawn for irrigation (i) at time (t)	m^3
J	Total number of dams in the system	-
P_e	The economic benefit of generated energy	US$/MWh
P_w	The economic benefit of withdrawal water for irrigation	US$/m^3
$maxS_{t,j}$	Maximum storage capacity of reservoir (j) at time (t)	m^3/month
$Qin_{t,j}$	Total inflow of reservoir (j) at time (t).	m^3/month
$Qout_{t,j}$	Total outflow of reservoir (j) at time (t).	m^3/month
$q^{min}_{t,j}$	Minimum turbine discharge of reservoir (j) at time (t)	m^3/month
$q^{max}_{t,j}$	Maximum turbine discharge of reservoir (j) at time (t)	m^3/month

164

Symbol	Parameter description	Value/ unit
$R_{t,j}$	Release state variables from reservoir (j) at time (t)	m³/month
$Rw_{t,j}$	Releases through the turbines from reservoir (j) at time (t)	m³/month
$Rs_{t,j}$	Releases for sediment flushing from reservoir (j) at time (t)	m³/month
$S_{t,j}$	Storage state variable of reservoir (j) at time (t)	m³/month
S_j^{min}	Minimum storage volume of reservoir (j)	m³
S_j^{max}	Maximum storage volume of reservoir (j)	m³
$S_{t,j}^{max}$	The maximum storage capacity of reservoir (j) at time (t)	m³/month
$SDR_{t,j}$	Sediment deposition of reservoir(j) at time (t)	m³/month
$STR_{t,j}$	Storage-capacity ratio of reservoir(j) at time (t)	-
$SR_{t,j}$	Sediment released from reservoir (j) at time (t),	m³/month
$Sp_{t,j}$	Spillage of reservoir (j) at time (t)	m³/month
T	Planning time horizon	month
$TE_{t,j}$	Trap efficiency of reservoir (j) at time (t)	-
$\tau_{t,j}$	Number of hours in period (t)	hours/month
$Y_{t,j}$	the sediment yiled of the reservoir (j) at time (t)	tons/month
$\eta_{t,j}$	Turbine efficiency	-
φ	Sediment dry bulk density	ton/m³
$\propto$	Coefficient representing supply/demand ratio	-
$\omega_1, \omega_2, \omega_3$	Weight factors of the respective objective functions and satisfy the condition	-

LIST OF TABLES

LIST OF FIGURES

ABOUT THE AUTHOR

Eng. Reem Fikri Mohamed Osman Digna is a lecturer of water resource engineering at the Faculty of Engineering, University of Khartoum (UofK). Her research work focuses on mathematical models for decision making in support of water resources planning and management, urban flood analysis, and water distribution systems simulation and optimization.

Reem graduated from Civil Engineering with First Class Honor from UofK in 2002. She worked as teaching assistant at the same university, conducting tutorials and organizing laboratory work. Following her MSc degree in water resources engineering from the University of Khartoum in 2008, she was appointed lecturer, teaching undergraduate courses of hydrology, fluid mechanics, and hydraulics, and the graduate course of numerical methods and application in water resources modeling. She is a member of Water Research Center (WRC) at UofK, where she also is the focal person of the Nile Basin Decision Support System Stakeholders' Community.

In 2012 Reem joined UNESCO-IHE (now IHE Delft) as PhD fellow with a research focused on optimizing the Eastern Nile multiple reservoir system. During her PhD, she participated in several international and regional conferences in Bangladesh, Egypt, Ethiopia, the Netherlands and Canada. She supervised a couple of BSc theses. She worked as coordinator (2013) for the Water Resources and Environmental Engineering MSc program. She also published in national and international journals.

Reem enjoys spending time with her family, outdoor activities and volunteering in community services. In 2018, she was awarded exceptional service award by Etobicoke Center in Toronto, Canada in recognition of her contribution to the betterment of community.

Journals publications

Digna, R.F., Castro-Gama, M., Zaag, P.v.d., Mohamed, Y.A., Corzo, G., and Uhlenbrook, S., 2018. Optimal Operation of the Eastern Nile system using Genetic Algorithm, and Benefit Distribution of Water Resources Development. Water Journal, 10(7), 921.

Digna, R.F., Mohamed, Y.A., Zaag, P.v.d., Uhlenbrook, S., Krogt, W.v.d.and Corzo, G., 2018. Impact of Water Resources Development on Water Availability for Hydropower Production and Irrigated Agriculture of the Eastern Nile Basin. Journal of Water Resources Planning and Management, 144(5): 05018007.

Digna, R.F., Mohamed, Y.A., van der Zaag, P., Uhlenbrook, S.and Corzo, G.A., 2017. Nile River Basin modelling for water resources management – a literature review. International Journal of River Basin Management, 15(1): 39-52.

Ahmed, M.I. and Digna, R.F. Stormwater Quality and Pollutant Accumulation Rates in the Surfaces of Khartoum State, 2007. Journal of the Sudanese Engineering Society, 53(49):33-45.

Conference proceedings

Reem F. Digna, Pieter van der Zaag, Yasir A. Mohamed, S.Uhlenbrook, Mustafa A Mukhtar, " Optimizing reservoir operation to include sediment management: a new approach tested for the Roseires dam, the Blue Nile River", CHI International Conference on Water Management Modeling, Toronto, Canada, 27-28 February 2019.

Reem F. Digna, Wil van der Krogt, Yasir A. Mohamed, Pieter van der Zaag, S.Uhlenbrook, " The implication of new dams construction on the Eastern Nile water availability for hydropower and irrigation at regional and national levels", Eastern Nile Basin Hydrology and Climatology Scientific meeting, Cairo, Egypt, 19-20 April 2017.

Reem F. Digna, Wil van der Krogt, Yasir A. Mohamed, Pieter van der Zaag, S.Uhlenbrook, " The impact of Grand Ethiopian Renaissance Dam on the hydropower in Sudan and Egypt", Water and Electrical Energy and water desalination Conference-2016, Arab Engineers Union , Aldoha, Qatar, 22-23 February2016.

Reem F. Digna, Wil van der Krogt, Yasir A. Mohamed, Pieter van der Zaag, S.Uhlenbrook, " The implication of upstream water development on downstream river basin: the case of the Blue Nile River Basin", the 7th Graduate Studies and Scientific Research Conference-2016, University of Khartoum, Khartoum, Sudan, 20-23 February 2016.

Reem F. Digna, Wil van der Krogt, Yasir A. Mohamed, Pieter van der Zaag, S.Uhlenbrook, "The impact of new dam construction on the Eastern Nile water availability for hydropower and irrigation at regional and national levels", The 5th International Conference on Water and Flood Management (ICWFM-2015), Dhaka, Bangladesh, 6-8 March 2015.

Reem F. Digna, Wil van der Krogt, Yasir A. Mohamed, Pieter van der Zaag, S.Uhlenbrook", The impact of Settit and Grand Ethiopian Renaissance new dams on the Eastern Nile River basin,' The Second New Nile Conference, UN-ECA, Addis Ababa, Ethiopia, 8-9 December 2014

Ahmed, M.I. and Digna, R.F., "Urban Runoff in Khartoum and its Effect on the Nile Water Quality", V Middle East Regional Conference on Civil Engineering Technology and V International Symposium on Environmental Hydrology, ASCE-EG, ESIE, Cairo, Egypt, 3-5 September 2007.